BATTLE OF AMSTERDAM

by Rob Hackney

Battle of Amsterdam

ISBN 978-0-9875370-0-3

Cover design by JJ Wind, Rob Hackney and Andrea L.
Book Design & Typesetting by Andrea L.

First Edition.

Disclaimer

All of the people, places, and especially the corporations named in this book are entirely fictional, even those based on contemporary, existing brands. The geopolitical situation described herein represents an imagining of the players and scenarios that could one day come to pass, and is in no way meant to imply anything negative about any specific modern day, real world organization.

For Anne, Snickers, and Mum & Dad

Also by Rob Hackney

TEMPUS CIRCA

ERASMUS

RANGER

PROLOGUE

last train out of Iceland

Otis Wilmington is being followed.

They were there with him in Prague, two days ago. In London, and Monaco. Now they're with Otis on the platform of a seventieth floor light rail terminal, waiting amongst the crowd.

For three days and nights they've shadowed his movement across continents.

Soon they will close in, snatching up the key that was carried illegally through borders and strong points these past seven days. A week is all it ever takes. The exchange won't happen on the train, or the next shuttle service, but in an airport hotel thirty to sixty floors beneath the Singapore tarmac. Quiet and e-shielded. Privacy, if not anonymity.

Otis knows for certain that he's in no danger. He is well aware that his unseen entourage will do anything necessary to ensure the integrity of data he holds, even if it means storming a checkpoint on his behalf.

Despite the time and cost involved, Otis isn't anxious. He has done all of this before. Will do so again, and again, probably for the rest of his life.

far from home

Ennis Pryce can't return to Amsterdam, under penalty of death.

From the streets of Rotterdam he watches the tightly guarded bridges and river, the sentries atop colossal dividing walls, patrolling for those who would try entering the utopian isolationist artificial island.

From high points in the Rotterdam CBD he catches glimpses of daily goings-on amongst the visible parks and streets, trying to hold the mental picture of calmly perpetual summer days, spent wandering around with no real aim or burning ambition.

For six years Ennis has been trying to go home. Every day since leaving he has found new ways to fail at repatriating, each avenue all but exhausted now. His will and overall health deteriorate more rapidly despite the growing notion that there is within him a grand design still undiscovered. Although he can't yet chip away at its edges he senses the scale of it, understanding that it will be both more final and more decisive than what has come previously.

Cajoling, bartering, both wheeling and dealing. All fruitless. What he hasn't done thus far, he reasons, is make a threat. Really threaten the shit out of those fuckers, then slip in unnoticed while they're all cowering in bunkers.

He tries to imagine the kind of threat that could make the notoriously stubborn alliance holdout acquiesce.

And then, all at once, there it is. Surprisingly obvious. Deceptively simple. Slightly mental? Of course. But all great ideas seem a bit that way at first.

end times

Imogen Harper feels tired all the time.

She wonders if being asleep more often than awake is a fated preparation to leave the physical realm. The anticipation of what those first few moments may be like, the sensations experienced, is never far from her mind. It consumes her now moreso than fatigue.

Try as she might, Imogen can't imagine how an afterlife would look or feel. Would she lose the gamut of sensory perception, or would her level of experience suddenly widen?

The thing that she would miss most about this particular life, really the one and only thing she can nail down, is curling up under a duvet on cold mornings.

If nothing else, she hopes it would be cold there. But what of blankets?

intervention

Umberto Daily has no idea how much shit he's in.

He eases back on the thrust as his medevac helo hits Mars atmo, deploying avionics. The cabin begins to shake here, a violent mix of turbulence and age.

While this trip to the Gusev desert is one of hundreds, it will easily be his most memorable. Coming in low between floating highway markers, still a thousand clicks from any city, Umberto can see smoke on the horizon.

dangerous

Apple Zenith searches her interactions with people for a tangible purpose. Obsesses over the meaning of gestures and brief flashes of sentiment that seem to ripple across faces like waves of neon, telegraphing intent. She has difficulty reconciling it with the words that leave their mouths, and so avoids looking at people when they speak.

Mostly she struggles with the overall definition, or line between, something that matters and the other. The daughter of an insurance CFO and a World Gallery artist, she will be taken care of for life. Nothing that she could ever want on this world or any must be sought out. Things are perfect here. Ease, solitude, a lack of the fear that's mostly evident in those below. How to pay for something. Getting to work on time. Responding with the appropriate intonation and level of enthusiasm, so as to secure continued social relevance.

People see her as a cold fish, is what she gets from the little she has read about herself. This much is probably true. Someone who has forgotten how to emote is surely something of an ice cold bitch whether or not she intends it that way.

Insulated from humanity at large by her penthouse home in the sky, the hardest thing in the world for Apple is to be redundant.

And while she is aware that there is another way to be, that it is furthermore attainable, Apple could not even begin to guess the first and second steps to achieving it.

One

killing time

.Singapore Transit.MoD/AP08—Asia.2152.11.11.04:12

In a cramped, filing cabinet hotel room thirty floors beneath the continental airport tarmac, Otis leans over a small basin, scooping warm water to wash his face and chest.

The lease expires in three hours, forty-something minutes. People sleep here, four hours at a time. For them an opportunity to shit, shower, shave, pace and recline, before continuing to a connecting flight.

For him it's an office. Tainted with the indescribable sensation of the place where you work. Between handoff rooms, more hotels just like it. Sometimes bigger, nicer, worse, smaller. They're all his work, not anything like a home. Never home. There isn't a place for Otis to truly unwind. Sure he can shit, and all the other ones. But even if he went on holiday, he would still find himself in some similarly fouled hotel room.

His life, metered in units of four, is indistinguishable from episodes of a television show that never really seems to go anywhere. He feels like the archetypal cartoon character, frozen in a looped kind of continuity for years on end, reliving the same basic adventure with slightly different personalities and backdrops depending on the week.

audience with the board

.Seoul Anchorage.MoD/SD01—Asia.2152.11.11.04:19

Anchored just outside In' Chon harbor's intercorporate shipping lanes, the McPherson Group's aerophibeous half-kilometer yacht rocks gently in a growing tidal swell.

"We lost them," says McPherson Group's Head of Comms, registering the worried looks before adding, "we'll have ULF up in a moment."

A bureaucratic murmur, then an incoming call alert, and finally a content hush from the decision makers aboard McPherson's second largest yacht.

"We're on approach," his field based sub reports to the room.

A display high on the wall reads, 'Audio Only'.

"Get that port ready," says one of the board members in back, nameless here under operational guidelines.

An analyst who mans feed stations in the corner announces, "It's open, prompting for hash check."

data dump

Otis lets the executives in, watching their comms tech privacy scan the small room, while another lays a PDA down on the bed.

As Otis retrieves his own from the coat pocket, the third man steps into the bathroom, activating the shower's tanning lights then several of the handheld privacy crawlers worn on his belt.

Otis places his PDA beside the other and taps its lightpad several times, navigating quickly through menus. The display reads, 'Pairing . . . ' then updates with 'Hashtag sent.'

Otis tilts the projected confirmation toward the execs.

The comms guy nods as he listens to his earpiece, looking up at his colleagues with apparent relief. "We're good."

The second suit shows Otis a deposit receipt for €910,100.

we came all this way

The transmitted encryption key is accepted by the analyst's terminal, displaying a previously inaccessible set of menus on targeting hardware purchased through loopholes in the trade embargo on Sun Microsystems.

The room's focal processing station hums to life as another nameless higher-up broadcasts, "Wilmington's off tilt for the month. See him out."

thanks

Otis leaves the room with a restrained nod to the others, swiping his PDA past the checkout display. As the rest pack their gear, a digital voice announces that they have forty seconds to vacate.

In the elevator, ascending fast to the terminal, Otis asks himself if he wants to get drunk tonight, already sure of the answer.

For as long as it takes the doors to glide open, he wonders if he will spend the rest of his life in airports.

border patrol

.Rotterdam Residential.PP/EN12—EU.2153.03.24.20:50

From an apartment halfway up Sky City Seven, facing north amongst employee housing in Paypal's Neighborhood Twelve, Ennis leans from the narrow window carefully, still draped in the dark gray sheet with a monocular protruding from beneath its folds.

He comes back to record the time then moves to another window, repeating the exercise.

Three clicks away border guards thoroughly inspect parked rail cars by the southern entrance to Amsterdam's freight tunnel.

Graffiti on the outer concrete wall reads, 'Everything has its place in the heart of man, for even what he fears defines him…'

Below it, 'Fuck off cuntorrrr!!!'

inside the fold

.Amsterdam Sovereign Economic Zone—EU.2153.03.26.14:08

Long steamers drift silently through Amsterdam's CBD two wide in its narrow canals.

Several floors up, on the balcony of an exclusive coffee shop, Serrah Zenith paints the scene below.

PDA vibrating softly, she takes a moment to finish the detail at one corner of a digital canvas.

"Accept it," says Serrah, smoking a blend of coffee and herb. "Hi hi."

Her daughter's face flickers on its projected display.

Apple says, "I'm leaving college. I thought you should know."

"Oh, that sounds fun."

"Take me seriously."

"I never presumed to do otherwise, dear."

Apple pauses. Serrah tilts the panel toward herself, unable to read her youngest after nineteen years.

Serrah asks, "Why don't you come for a visit?"

"Negro please. I'd rather work for McPherson."

"Well I do appreciate your calls."

ace of sleeves

.Rotterdam Transit.PP/MN03—EU.2153.05.06.22:14

Three blocks from the central hub of PayPal's mag-lev network connecting Target and KrugerIS regions, the train makes its final seventieth floor stop.

"I'm basically looking for a bomb," Ennis tells the stranger on call. His PDA reads, 'Audio Only'.

"Okay?" prods the arms dealer.

Ennis looks around the carriage, lowering his voice some.

"It's gonna have to be nuclear."

"Regular, or large?"

"Oh, I hadn't really thought about it. What do you recommend?"

"Well, you know what they say about, ah, it's better to need a regular and have the large, than need a large and have the regular."

"Yeah I heard that somewhere," says Ennis. "Seems like sound advice."

"It'll cost you though. More than the composition sticks. And because I don't know you, I need it all upfront."

"I'm working on it. We should be good."

Ennis pockets his PDA, vacantly staring at floating advertisements while the train descends rattling into Rotterdam's neglected subway system.

tinfoil sombrero

.Eindhoven SubTransit.KI/UW56—EU.2153.05.14.12:03

"What's this one?" Ennis asks the man who for some reason wears sunglasses underground.

"Fission device. One gigaton."

"Okay this may seem like a strange question, but which one puts out the most passive radiation?"

The dealer shakes his head. "You don't want that. Meteorology would register it on the flyby, they'll stumble onto it in a matter of days."

"What if I had a timetable?"

"On company satellites? Good luck breaking the encryption. Trust me, it was my job once."

"Just humor me," says Ennis.

"Okay, well . . . I guess you'd probably be looking at something pre- twenty-fifty. Neutron bomb, ninety megaton range."

Ennis slowly nods, eyes narrowed.

The dealer seems concerned by this. "But you'd want to be damn sure you know when those birds are overhead before taking it out of the bunker. Peace

Enforcement don't patrol under the highways anymore, the people who have to work here really like it that way."

"Next time you see me," says Ennis, "I'll have the cash and flightplans. For now though, I'll just take this sweet revolver."

Ennis taps his PDA, which chimes as he waves it over the dealer's home feed station.

The arms dealer lights a cigarette and clears his throat, returning all attention to the muted feed display floating above his bomb shelter's kitchen sink. "Volume," he tells it, filling the cramped bunker with synth music and excited Hindi.

Ennis leaves the room past shelves stocked with tinned food and bottled water, ascending a stainless steel ladder back to the surface.

He climbs out of an alloy blast door in the back yard of the arms dealer's ghetto safehouse, leaving via a side gate. Despite the midday visit, no sunlight penetrates the layers of concrete highway overhead.

Large, pale eyes watch Ennis return to his idling taxi, which peels away from the curb the moment he's inside.

Riding in back, Ennis keeps one hand in his pocket, feeling the revolver's shape and weight. For as long as it takes the lights to change, he wonders if it will feel strange to use it on a man.

time sink

.Montana Residential.LM/EH19—North America.2153.01.09.21:44

Imogen slept all day again, getting up only when she hears the garbage collection pull away from her small townhouse, the fading rumble ubiquitous in a quiet street of architects and engineers.

After being awake just an hour she's finding it hard to maintain awareness. Slipping in and out while watching feed dramas, catching thirty seconds of plot for ten minutes' nap.

She started getting sick around the same time as her husband, but mostly let it go untreated. Chronic fatigue hardly seems noteworthy having been around chronic pain. Imogen knows that there are worse things than being tired.

A strong coffee sits cold on the table within arm's reach but untouched.

On feed, bleeding in and out of the dream, someone else's love goes awry.

white glow

In the dairy aisle of an all night market Imogen struggles to read the expiration date on a liter of milk, eyes burning.

She approaches the bored kid behind bulletproof glass up front. "Can you tell me when this milk expires? I don't have my lenses."

He takes it from her through the slot, glancing at its label before swiping the carton.

"You've got a week. It's the seventeenth."

"And a pack of Guarana filters."

"We're out of them."

"Ah, just give me some caf pills and a coffee-flavored Wrigley's then."

spread thin

"What kind of shit is that?" Imogen demands of her home feed station, voice strained and cracking. She taps *Vector* again.

'Error in field *Physical Value [Gr43879hj.tar.kru]'.

She sips espresso, still glaring at the floating display. After scrolling up, re-checking the columns, she tries it once more but knows it can only error. Whatever snag she's hit within the code will wait until morning. Or afternoon perhaps. Whenever she can bring herself back to it.

"Log out," she tells it, slowly getting to her feet. "Infotainment, local news."

Imogen deflates into the couch, watching the news a moment before drifting off.

gray skies

.Unpartitioned Gusev desert—Mars.2167.09.02.10:31

1, 415 clicks outside Gusev City, heading south-south-east, he sees it at first as a smudge, but knows right away that this is his wreck.

Umberto passes the downed freighter's impact point at half-throttle, following the slight plateau's terrain around its wide arc, visually assessing the damage from his helo's cabin while every hotrigged surveillance mod aboard scans ineffectively for signs of life.

The charred debris is smeared across eight clicks of red, mostly flat desert. A smoldering stain on once scenic landscape.

"Tokyo?" he says into the live headset, "I don't know what kind of beacon you think you're reading there but I can assure you, it isn't a pleasure craft."

Having come across this kind of scenario twice before in his short career, Umberto knows that it will be bloody. He can't imagine finding any less than fifty crew on a ship this large.

"Guys I'm gonna go down there and assess the impact, you might want to standby a Pelican and alert Gusev Region Memorial."

budgetary setbacks

The Helo's port door slides open with a hiss, letting in ozone-scented Mars air.

Umberto steps outside, PDA scanning the wreckage for crash tube locaters.

"We're sending you imaging now," says the comms man back in Tokyo, tinny and distant in the earpiece. "Try walking east about a half click, there's a huge chunk of something there. Our analyst said it's the bridge, but one of the ship's architects seems to think it's crew quarters, at a glance. He's getting back to us."

Umberto treads through the ashen mud, toward more towering masses of smoking wreckage.

"If you guys would spring for the search and rescue package, I could do most of this from the air."

"Hey don't bitch at me," snaps the comms guy, "I'm still waiting for them to buy me a chair. I do this job sitting in a window frame."

Umberto begins to scan the bigger pieces as he passes. "Okay Steve, you have it tough too. Jeez."

The PDA flashes a low-level radiation warning.

"Hey Tokyo this stuff is reading kinda hot. Should I be wearing a suit or something?"

"Ahhh . . . Tokyo will get back to you on that."

life raft

"Architect nailed it," Umberto tells comms. "Crew housing."

He makes his way down the steeply-sloped interior corridor, rubber shoes gripping tentative on stainless steel.

A tungsten strip on the front of his flight suit glows harsh white, lighting the way ahead. In his hand, the PDA flashes an alert.

"Tokyo, I'm getting an activated crash tube still in its rack."

Umberto turns left through torn open alloy wreck doors, into a large room lined by suspension-mounted cylinders.

A display on one of the pods strobes blue-green-red.

He goes to it, pulling the release on a panel marked 'Rescue', taking a couple of tentative breaths as he waits for it to respond. He balks at the thought of having to hoof it back to his helo for the jaws of life.

Finally the front half of the crash tube shielding slides forward before rolling open, revealing a woman enveloped in translucent blue gel, sealed behind plexiglass.

She takes short, anxious breaths from a small o2 tank.

Conscious but frightened, her eyes dart about adjusting to the light, before settling on Umberto.

He punches the flashing red disc and its safety glass slides away, releasing the thick goo over his suit from the waste down, the girl collapsing in his arms.

Umberto drags her clear and lays her down coma style, removing the tank's mouthpiece, wiping gel and wet hair from her face.

"Can you hear me?" he asks. "Are you hurt?"

"We can't stay here," she tells Umberto weakly, holding her throat, eyes wide and blinking fast.

"It's okay, I'm with an evac firm based out of Asia. I'm taking you to Isidis for medical care."

"Not there," she says. "Nowhere on Mars is safe."

"I don't understand."

"There aren't meant to be survivors. The alliance will come and hunt me down. Probably you too now."

"You're in shock. You've been in a pretty bad accident."

The woman shakes her head, fighting back tears.

"We have to *go*," she says with an urgency that Umberto rarely hears. "Please! We don't have time for you not to understand what's happening."

"If I take you off world our billing department will go to town on your insurance."

"Sir . . . if you don't get me out of here in the next ten to fifteen minutes, you and I are likely to be found amongst the next crashed freighter."

fallout

.Ontario Commercial.MG/BD01—North America.2154.06.20.12:39

Oblivious to the sobbing girl behind her, the shocked gasps and panicked murmurs in the room, she fixates on the feed display without taking in any of it.

For now Apple can only think about how this is the place she will remember being when someone inevitably asks the question, "Where were you when you heard about Amsterdam?"

How humiliating.

Apple finishes her lunch quietly and leaves the break room, making her way past ashen men and women who stand silent in cubicles, eyes glued to the feed.

She enters her small office and shuts the door, stopping by the narrow window.

Air traffic outside has dried up, while cars below have turned the streets into a parking lot, drivers grouping around larger screens.

The world is suddenly different now. This is something new. Something you wait a lifetime to see, always hoping you won't.

Still, she wonders why she doesn't feel any different.

Projected by the oak desk, her display reads, '410 new Q-Mails, 409 marked urgent'.

Apple sits down, placing her PDA on the varnish.

"Call mom," she calmly tells it.

A digital voice says, "The subscriber you are attempting to call is not logged in."

"Call dad."

Her father's face appears onscreen almost instantly.

"Apple . . . "

"Is she okay?"

"I don't know. We have people in Rotterdam, but they can't get through the road blocks."

"Dad . . . is it really that bad?"

"They think so."

An uncomfortable silence.

"Apple I have a lot going on here. As soon as I know anything, so will you."

"End," she says, leaning back in the leather chair.

A knock at the door then an analyst appears there—nose red, mascara running.

"Oh God . . . Ms. Zenith," she says between sobs, "did you hear?"

For some reason Apple shakes her head, offering the woman only a blank stare.

"Someone dropped a nuclear bomb on Amsterdam."

"Oh no. Will everyone be okay?"

"It's gone." The analyst pads away tears with a makeup stained handkerchief, struggling for composure. "They're saying it's all gone."

Two

Vacuum

.Intercorporate shipping space—near-Mars orbit.2167.09.02.10:31

At the tail end of exit trajectory, when gravity's pull has weakened to the point that loose items drift about the flight deck, Umberto engages autopilot and unbuckles his harness, returning to the rear cabin.

The woman sits up on one of two med station beds, strapped to it, sipping electrolytes.

"You should lay down," Umberto tells her, taking the bottle.

"I'm okay."

"Umberto!" yells his dispatch guy on speaker. "Is she aboard?"

"Comms down." He sits on the other bed, bracing himself, their knees almost touching.

The woman takes a moment to study his face now, the shock beginning to wear off. He looks about thirty, maybe half or a quarter Latino. Good skin . . . even at this distance, she can smell that he uses a Sorbol based moisturizer. Shaved a day and a half ago, meaning he doesn't hate his job yet, but is on the way there.

"Umberto..? That's you?"

"Yeah. You?"

"Monique." She breaks eye contact, looking around the small med-station.

"This ship, it's Military?"

"No, not a ship."

She shrugs. "What would you call it?"

"A ship uses chemical energy for thrust. Helos have magnets. Big, spinning magnets."

"Okay so is it Military?"

"Used to be. St. Martin's Ambulance buy them at auction. We lease their surplus fleet. Why?"

"And who are 'we'?"

"Medevac Asiatic."

"So you guys are like a private ambulance?"

"All ambulances are private."

"Not on Mars."

"Well Mars doesn't count."

She shakes her head a little. "What's the plan here?"

"Take you to my dispatch building, call you a cab, go home and sleep."

Panic flashes on her face. "Earth?"

"Yeah."

"We can't go in through Quarantine gates."

Umberto laughs. "Avoid Quarantine? The fine is ridiculous. Like, give me the prison time ridiculous."

She leans forward, placing a hand on his elbow.

"If my DNA goes on file at the gates the people looking for me will find and kill us both."

"Why would that happen? Are you some kind of spy?"

"I'm a Zenith."

"Zenith?"

"Yeah, like the insurance guy."

"Okay. How does someone like you end up on Mars?"

"Long story."

"How about over dinner?"

She frowns. "Are you hitting on me?"

Umberto peels the lid from a ration kit, handing it to her.

"No, we don't do that kind of thing. It's one of the main rules." He points to a charter taped to the dividing wall.

sovereign borders

.Quarantine Gate GE12—near-Earth orbit.2167.09.02.11:27

Bolton O'Hare returns from the smoking room deep in thought, finding a queued call at his desk.

"Answer," he says before coughing into his sleeve. "GE-12, you're speaking with Bolton."

On speaker, the caller says, "I need to speak with your manager."

"I'm the manager. This is he."

"They make you answer calls?"

"Quarantine's spread a little thin these days. How can I help you, sir?"

"Well that brings me to my point, Mr...?"

"O'Hare."

"Mr. O'Hare I'll be brief. An unlicensed EMT flying medevac hardware will soon attempt to circumvent inbound Earth Quarantine with a woman considered by several members of the Healthcom Alliance as a class one contamination risk."

"Wait, what's a class one? Is that the highest or the lowest?"

"The highest, Mr. O'Hare. It's shit creek without a ladder."

"Paddle?"

"It's shit creek."

"Right." Bolton searches the desk for his PDA, "Okay, uh, when is this supposed to be happening?"

"You have roughly ninety minutes."

"That's not a lot of time."

"No," says the caller, "not at all."

"Can you tell me where?"

"Flight path is predicted to hit your boundaries somewhere between GE-12 and FD-10. The firm I represent were tracking him out of Mars, but the pilot slipped surveillance on the biways."

"Wow, this is pretty full-on. Is there anything else I should know?"

"If they reach Earth the virus will spread faster than any outbreak you read about in high school."

"I didn't really go to—"

"It will spread very fast."

"I should probably call someone."

"Probably."

"What did you say your name was?"

The display reads, 'Call ended by caller'.

"Classy," says Bolton, then "call Naval Liaison."

dogfight

.Hyperspace biway ME04—Earth-Mars space.2167.09.02.12:06

Umberto's first inkling that something has gone wrong is the formation of interceptors showing up on long-range recon just a few clicks beyond the exit gate.

He turns to Monique, strapped into the co-pilot's chair.

"Thoughts?"

She shakes her head.

"Is that accurate? There can't be more than fifty."

"Fifty-nine," he says. "The data comes from three relays. So, yeah, it's about ninety-eight per cent accurate."

"Can we run?"

"This thing was built as troop transport. It doesn't do anything fancy." Umberto cycles through news and emergency services feeds, finding nothing on the sudden armada. After some thought he says, "Comms up."

"Umberto? What happened?" asks Steve.

"Hey what's the deal with Quarantine and Navy? I'm seeing a squadron and a half spread out between GE-13 and FD-09."

"It's up on the industry ticker dude, some unlicensed medico is trying to slip quarantine with an Ebola patient or something."

"Ebola?" Umberto throws a glance at Monique, who shakes her head. "Did they say who the guy is?"

"If they knew that, they'd probably just go straight to his firm."

"Right . . ."

"So what happened with the Mars chick? Was she hot? I'm still waiting for the check-in from Isidis. Did you hear about Carlos's burn victim at Royal Elysium, took them four days to send the check-in, guy was dead by the time we billed him."

"Yeah . . . fuck those guys. Listen comms has been in and out since I took off, might want to book this bird for service."

Umberto shuts off comms via the lightpad.

"It's us right?"

"Yeah."

"You're infected with something?"

"When an employee turns whistleblower they tell Press and Quarantine that they contracted something hazardous from a biotech lab and need to be returned to specialty isolation—within company territory of course. The sham virus always seems to get the better of them in the end. Never an autopsy. Never an inquiry. It's all sealed."

Umberto frowns, unsure. "What kind of twisted-ass company pulls shit like that?"

"The biggest."

"Next to McPherson maybe."

Monique says nothing.

"You're saying it's *McPherson?*"

"I'm not *not* saying it."

"Balls." Umberto shrugs, tossing up his hands. "Game over then."

"This is what I've been trying to get across. We can't just let them intercept us. I'm sorry, but you're in it too now."

"Okay . . . " His eyes dart about the cabin, checking timeline and course projections. "Here's what we do . . . they're looking for a Medevac helo *avoiding* Quarantine, right?"

"Uh huh."

He licks his lips, brow furrowed. "Going back and forth over the years . . . you figure out which gates are older. Which ones have the sketchier equipment. We sneak the occasional liquor stash through Quarantine, but never people. You might want to go in back and start making a little fort out of anything with carbon."

a straight line

.Quarantine Gate AA02—near-Earth orbit.2167.09.02.12:52

The UH-60 rolls lazily through the gate's three inner rings, which rotate as they scan the ship from nose to tail.

"Anything to declare?" comes the Quarantine guy on comms.

Umberto leans forward and smiles at the man in the window. "Sorry, no, I don't have any witty statements like 'don't go to space' or 'yeah, my wife is a bitch' for you today. Maybe next time, guy."

Barn doors swing open on the third ring, revealing blinding sunlight for the instant it takes the windshield to adjust.

The helo lurches forward as its internal magnets spool up. It exits the station and sets course for Asia, glowing orange with the first sniff of atmosphere.

ticket to war

.Antwerpen Commercial.Departition—EU.2153.08.18.23:51

Ennis waits in the lone standing doorway of a collapsed firehouse, concealed by its shadow, watching the destitute and miscreants shuffle by on their Friday night chemical hunts.

One of the migrating homeless stops when he spots Ennis there in the dark, pulling back his hood to reveal a clean-shaven face.

A hint of dress shirt collar and striped Windsor knot are visible beneath his torn sweater.

The faux hobo shows him a coffee mug, rolled-up envelope stuffed into it.

"Do I . . . take it?" Ennis asks.

The messenger sighs, cautiously looking up and down the block before returning his attention to Ennis.

"What are you retarded?"

Ennis snatches the envelope, pocketing it as the guy moves on.

He waits half a minute then walks the other way, snaking through the open-air drug markets of lower Rotterdam.

the job

.Berlin Admin.YM/HR04—EU.2153.08.21.09:16

The suit he wears is worth €99 but it could easily pass as lower-end brand name. It folds and grabs in all the wrong places, itching him of course, but Ennis remains motionless in the chair.

Behind his desk, Pico Tennerman reads over the CV on display, scanning for points of interest. Behind Pico the city sprawl, an impressive eightieth floor view marred only by the occasional uncomfortable meeting of eyes with pilots as they pass at a crawl in lanes just beyond the glass. Yamaha's Four district is notorious for both its unreasonable traffic and hiring policies, neither of which should hamper Ennis.

"I see that you studied in Ontatrio," says Tennerman.

"No, I studied Ontario. For a while. Read about things that happened there. Things that were near it. French Canada, for example, was in the nation of Canada, not France. A lot of people might not know that." Ennis's eyes dart again to the window.

Tennerman chuckles, unable to really hide his discomfort. "Well, as long as you're not a Maple Leafs supporter."

"Who?"

"Uh, they're a hockey team."

"Laser hockey?"

Tennerman frowns. "*Ice* . . . hockey."

Outside, an automated washing rig rises into view then hovers in place, mechanized scrubbers tending to Tennerman's floor-to-ceiling windows.

"So . . ." Ennis leans forward, hands flat to the desk, "do I get the job or what?"

"I'll be honest with you Mr. Pryce, nothing on your CV stands out. You're not quite *wowing* me. Why should Yamaha hire a guy with very little experience outside of Amsterdam to liaise with MoD?"

The maintenance platform slams into Tennerman's window, startling the executive. He turns to watch it back up, readjust.

"Whoa! I thought it was a bird, those rigs are supposed to—"

It hits again, this time breaking the safety glass, plowing through it, pushing into Tennerman's high-back leather chair as he leaps up and scoots behind Ennis, the shattered glass collapsing as a sheet while the rig keeps coming, settling against the desk before finally powering down.

"Well that's weird," says Ennis, also getting to his feet. He walks around the desk, leaning over the errant machine.

"Can you see what went wrong?" asks Tennerman.

Ennis kneels beside it, tearing duct tape from the engine housing. "Not quite. There's something taped to it though."

"Is—wait! It could be a bomb?" Tennerman takes another cautious step backward.

Ennis rises from behind the desk with a revolver leveled at his one time former potential employer Pico Tennerman.

"What kind of shit is this?" asks the nervous Tennerman, half-raising his hands, putting it all together. "How did you get that past Security? You . . . what? Planned all this?"

"Sorry, but you're not exactly wowing me either, Bob."

"Pico."

Ennis shoots Pico in the neck, then crosses the room to put another round in his chest.

While Tennerman bleeds out on the floor Ennis tosses his gun through the window, exiting via the main door.

Walking to a bank of elevators Ennis tells the secretary, "He said give it ten minutes before the next guy."

He anticipated the shakes, blurred vision, jelly legs. Read about every potential physiological response. But there's none of that. He doesn't even feel any different than before. Ennis could walk back in there and do the same to everyone on the floor and all it would matter is that he might be late for his train.

blind spot

.Eindhoven Subtransit.KI/UW02—EU.2154.06.20.08:03

Ennis waits in the hired utility a block behind where uberway ends and sky begins.

In its cradle on the dash, his PDA counts backward from 74 seconds. His eyes slowly adjust to the light.

When the counter hits zero he floors it, roaring out of the projects and back to society. The PDA counts down again, 119 seconds.

He leans forward over the steering column, noticing gray clouds roll across a horizon bloated by early morning industry.

Rattling about in back, secured hastily by chains, the ninety-eight megaton neutron bomb makes Ennis feel empowered for the first time in six years.

Today corporate might and private borders seem slightly less insurmountable to the man who aches for home enough to wipe it off the planet.

amsterdam day

.Ontario Residential.MG/SH02—North America.2155.06.20.18:31

Absently entering the 221st floor apartment that she has occupied for six months, Apple drops her PDA on the kitchen bench and opens the fridge, kicking off her work shoes.

"Infotainment, something light."

The display pops up in one corner of the adjoining living room. On feed a talk show returns from commercial. The host says to camera, "You're watching Vanderbolt. We're talking with Ennis Pryce, the man who gained notoriety when he leveled Amsterdam with a hundred or so megaton nuclear NATO warhead, one year ago today."

On B camera Ennis says, "That's a colorful introduction Kirk."

Apple closes the fridge, absently staring at the display. She's seen photos of Ennis Pryce, but never heard his voice.

His quiet, soft-spoken manner somehow surprises her.

The body language is unusual, even for a talk show guest.

To Apple, Ennis seems almost . . . whimsical. At ease with himself, and his place in the universe.

This man—rather, genocidal maniac—intrigues her.

"Lobbyists for the extreme left continue to push for an Ombudsman sanctioned trial," says the host, "while unionists and universalists alike insist that Amsterdam be classified a rogue state, posthumously as it were."

Ennis cocks his head, looking somewhere off-camera. "Let's be clear . . . Amsterdam was no Islamiyah. Amsterdam as an entity may have rejected the unilateral ambitions of Ombudsman, MoD . . . all the corporate unionists and their lobbyist shell firms, but it never tried to push that ideology beyond its borders."

"True," says Kirk Vanderbolt.

"It's important that people out there understand *why* the whole Amsterdam thing happened."

Apple steps closer to the display.

"And . . . why is that?"

"Not to be admired by anyone . . . "

For the first time since her mother died, she feels something other than cold.

"It happened," says Ennis, "because a misguided piece of shit who clearly should never have been given nuclear capabilities . . . in an unfocused moment clouded by *bitter* disenchantment . . . pushed the Instant Revenge Button in his hand. And, as an afterthought, ended the lives of more than a million people."

"Are you a . . . religious man?" Kirk asks.

"I was raised Catholic. I've lapsed, of course. You kind of have to."

"And if it's not too personal . . . geolegality of your actions aside, how do you ask God to forgive you for all those deaths?"

Ennis looks silently at the floor, eyes hidden by the hard lights.

"You don't. Mortal sin is a binary thing. You've either killed a guy, or you haven't. What's another million at that point? I killed my first as part of the whole convoluted effort to get back through Amsterdam's wall. I felt at the time like it would seal my status as a . . . not man, but warrior, or something. This idea I always had that I was generally quite peaceful, unless pushed into a corner. Every day I had an inkling of the capability that lurked there under the right set of keyturns, and part of me wanted to see it in daylight just once. When I stood in Pico Tennerman's office and shot him in the throat, I watched a guy die in front of my gun and then didn't feel ashamed, but relieved. Because I knew then that I could do all of it. And that in the event they tried to dick me aroud, I would convince them I'd burn their city to the ground. Even though, as I said right after, I had no intention of really doing it. I just could. So I killed this guy Pico, whose only crime was that he was encroaching on a contract that—a particular organization wanted quite severely."

Apple gags on the warm alkaline taste.

"I'd lost all sense of perspective," says Ennis looking at camera, at her. "I spent six years obsessed and let it lead me down the one-way path."

She embraces rage a moment, trying it on, only caged in her trademark restraint. Apple lets the sensation fuel a fire that rises and falls like an urge to both laugh and cry, nausea while it subsides.

"Then again," says Ennis on the feed, "maybe that's just some fucking bullshit anyway."

"What's the one thing you would hope the world at large could learn from your experience?"

"Don't blow anything up without a damn good reason. Maybe even, just don't blow shit up. At least sleep on it. Don't kill people, it's naïve, it's always a dick move, even in war. Because what you think is war, most other people think of as assholes running in and shooting up their life and all their stuff and they maybe want to just go on living, not having to worry about which of their neighbors all hate each other. I can tell you, Kirk. In the year since this happened, I've heard some fucked up things. People have tried to get me involved in sick, sociopolitical shit that has no reason for existing other than to tip the scales further toward all-encompassing dominance of the universe."

"Uh, well that doesn't sound good, maybe next time we can go into detail on some of that."

What really pisses her off about this guy is that he thinks he got away with it. The smarmy bastard, he gets off on all of this attention. Despite his protests, Ennis Pryce enjoys being the man who killed all those people on a whim and walked away with his balls.

She can't believe someone hasn't already tried to take him out.

posse recruitment initiative phase A

.Seattle Operations.GP/OP05—North America.2155.07.10.19:40

She sits in the back letting them read their minutes and get their localist grassroots political shit out of the way, patiently listening while four speakers share palm card wisdom with the crowd.

When the chairman calls out, "Any other business?" Apple stands, making her way downstairs to the front of their borrowed lecture theatre.

The chairman covers his podium mic, asking her, "What's your name?"

"Apple Zenith."

He looks at her for a long moment. "Seriously."

She sighs. "You wanna see some ID, mister?"

Into his mic the chairman says, "Apple Zenith everybody."

They all stare, most unmoved, some inquisitive. Apple takes the mic, adjusting it.

"Hi. Some of you may have heard . . . they blew up Amsterdam last year. Well I say 'they', but that's only because we all got so used to saying it that way. A lot of people don't believe that the one guy could be responsible for an atrocity like Amsterdam. They'll try to tell you he's a patsy for the larger conspiracy of One Worlders."

In the audience of ninety leftist Green/Peace employees, someone coughs.

"But the truth is . . . he did act alone. He was responsible for every part of the tragedy, from delivering the nuclear payload to pushing the button. MoD will never sanction the war trial of this man because unofficially, he's kind of their hero. He's now dining with their board and attending events memorializing the contributions Amsterdam made to the world of art and science. Proof to the world at large that mass murder can bump you into a higher social bracket. The man who destroyed a million families is untouchable now because of who he knows and what he did for them without even meaning to."

Apple takes a moment to compose, doing her best to suppress the emotion becoming evident in her tone.

"And so the world remains indifferent. Long before the region becomes inhabitable again, people will forget the way it all went down."

She takes a sip from the chairman's water.

"My mother died in Amsterdam, leaving me a hundred billion euro with no real focus or long-term goals until late last night. I now intend to use those resources to pursue Ennis Pryce's execution, with the aim of making him a public example—a footnote in the history books that will serve as a reminder to any future nutjobs thinking of nuking our cities that it's *not* fucking cool."

Apple writes her q-mail address on the darkboard before returning to the mic.

"If anyone wants to be a part of that, get in touch tomorrow morning. Thanks for your time."

She nods to the chairman then exits through a fire door, setting off the alarm.

bender

.Darwin Accommodation.HL/TR11—Australia.2152.11.18.13:43

A three star room in Hilton's public district must have been the best he could do while brownout drunk. Otis can taste three kinds of vomit. The throbbing skull would suggest scotch, but empty gin bottles say otherwise. Why would he even touch gin? That almost never happens. Things must have left the rails prematurely last night.

"Messages." He rolls over, head hung, eyes swollen.

Keo appears on display, saying, "Hey Otis, looks like you did good. Check continental."

"Feed," says Otis. "Continental news."

Nothing noteworthy. He tells it, "Volume up," then staggers to the bathroom.

When he returns the network screens amateur footage of a fiery object burning up on re-entry before splashing down at sea.

Over it an anchor says, "The orbital rail was one of nine in service by private armed forces. While it's not clear at this point where the attack originated, Peace Enforcement branches in several territories are co-operating with MoD in their ongoing investigation."

state of mind

"Think about it," says Keo on call, "you made—what, a million euro give or take? The guy who moved that cannon took home at *least* five-hundred mil. Probably more. Legally, everything's flexible with corporate backing. And morally? You and the arms dealer deliver two halves of the same product. You may take the four-continent detour to slip it by MoD, but don't think the dealers aren't just as handy with manipulating guidelines."

Otis shaves at the bathroom/kitchenette mirror, disinterested.

"Didn't you try to sell me on this last year?"

"You make the call and I'll have you set up in a *day*. Clients, suppliers, Sol-wide clearance. It's taken care of."

"What do you get out of it?"

"Ten per cent."

"Sounds reasonable I guess."

"Otis, think real hard here. Do you really want to spend the rest of your life in shitty hotels and delayed redeye shuttles?"

Otis washes his face and looks around the room.

flawed

.Montana Residential.LM/EH19—North America.2153.02.22.04:06

Three hours of troubleshooting have led Imogen to an impossible conclusion. "Call Lewing," she tells her workstation, leaning away to stretch.

Above it, Lewing Enright appears in his own window, hair disheveled.

"Imogen . . . ?"

"Sorry if I woke you."

"It's fine, what's wrong?"

"I'm at home auditing the Longhorn sims. There's something you need to look at."

"Simulations? This can't wait until morning?"

She glances at the clock. "Well it is morning. But no. I found a potentially fatal flaw in the BD-77's navigation code."

He takes a moment to sit up in bed, putting on lenses.

"Then your audit's most likely flawed."

"I assumed that too. So I booked some time on MIT-12 and ran it again. Same anomaly, every time. It's something we wouldn't normally see. A maneuver that's not part of any proving drills."

"Q-mail me everything you have, I'll be there in fifteen."

midnight oil

On Imogen's feed display, a simulated BD-77 Longhorn bomber maneuvers to break orbit around a digital planet.

"The simulated planet's mass," Imogen tells Lewing, "is the same as Mars. I give it a flight-plan . . . "

She feeds the sim-ship a set of co-ordinates, its nose beginning to shift away from the Mars-like body.

"Then revise those almost immediately . . . "

She punches in new co-ordinates. On display, thrusters along the port side fire, making the big ship groan as it struggles to correct its vector while pulling away from the planet, unable to maintain the plotted course.

"You can already see orbital integrity beginning to break down here . . . grav lensing forces navigation to correct, then overcorrect, buckling strong points nineteen through twenty-seven momentarily, losing its exit trajectory."

Lewing shakes his head, watching the display. "It happens. The nav kernel has multiple failsafes. You should see it swing main engines toward the planet and fire in about eight seconds."

"Yeah, it does. It works as intended, but only at first."

The ship's aft section falls toward the planet while the big engines at its base light, ramping up to full thrust, halting the fall entirely before slowly, the Longhorn begins lifting into space again, gathering momentum. Stress alerts appear on display as its hull warps along the backbone, almost at fracture point.

Leaning over the lightpad, she inputs a final dataset.

"This . . . is a sub-one hundred click crosswind at the stratospheric barrier." The ship shudders, beginning to swerve, unable to maintain control of its top heavy forward section. "Enough to throw it off course by four degrees."

Onscreen, the back end swings out as the nose rolls back planetside, the entire spaceframe doing three big barrel rolls before spiraling too tight, beyond the limits of its structural integrity.

The huge bomber suddenly twists itself in half, the front section falling back toward the planet on a rainbow of debris while its rear shoots off into orbit, disintegrating.

"Thrust and gravity rip it apart."

Lewing blinks. "This can't be right . . . "

"I've run it fifty-seven times."

"But then why hasn't it happened already? They've been in service more than three years."

"A long string of events need to line up for something like this to occur. But the point is that it can. And we didn't find it until now."

He sighs, massaging his jaw.

flagged

.MONTANA RESTRICTED.MG/PE00—NORTH AMERICA.2153.02.22.04:30

Eli Lylo receives the greenlight via signed q-mail at 04:30.

In nine minutes a helo will take him from nondescript office buildings inside McPherson Group's administrative district, across company lines onto Lauchlap Michael soil.

The brief doesn't explain why an affiliate corp's engineering team leader is scheduled for retirement. With most of these, the sender doesn't go into detail. Eli and his team act simply as proxy for faceless analysts.

His team enter the cage one at a time to collect weapons, each meeting Eli's gaze as they pass. He pockets his PDA and shoulders a rifle, checking the charge on its sights before leaving with the others.

waking up tired

Imogen sleeps on the couch, her back to Lewing, who sits up at her feed station still agonizing over data from MIT-12's physics engine.

He turns at the sound of Imogen's front door unlocking, aware that she lives alone.

"Uh, Imogen?"

The door slides open before five faceless and silent men in body armor enter the townhouse single file, taking up firing positions around the living room. Each soldeir aims a compact rifle at Lewing, who raises his hands, uncertain.

They only stare back.

Eli enters last, removing his faceplate and helmet. He taps the fifth man in, taking his place as the troop lowers his rifle to begin searching other rooms.

"It's okay," he says to Lewing, "we're Enforcement."

"Not Lauchlap Michael."

"Right."

His eyes dart to Imogen. "I thought they only use affiliates when someone really needs to go away."

"Sounds like a conspiracy theory."

"So . . . you're going to kill us, or . . . can I go back to work here?"

"She must trust you," says Eli, "to fall asleep with her boss in the house."

"It's not like that. She has a condition called chronic fatigue."

"I guess that means she'll sleep through just about anything." Eli nods to one of the others.

"Please—" says Lewing, interrupted by the silenced headshot. His limp body sprawls forward onto the rug, blood pooling beneath Imogen's couch.

The troop aims his weapon at her, waiting for Lylo to speak.

Eli walks away, stopping at the front door. The troops exit as they came in, quick and quiet

Outside, the helo's stealth mutes crickets and frogs, casting an eery silence over the block as it lifts off from a communal parking lot.

arrivals

.Jinagsu Airspace.MA/SW88—Asia.2167.09.02.15:56

Umberto can see his house from here.

After three hours waiting in a holding pattern over China their clearance is broadcast via orbital feed relays. The helo breaks its four click-high figure eight above dense urban sprawl and crosses Medevac Asiatic borders, headed for central dispatch.

"If we're not somehow one step ahead of them," says Monique, "then we're too far behind."

He checks regional weather and traffic tickers on the main screen. "Hopefully the first one."

Umberto levels out over an arterial highway, following its lights into town by eye despite the many nav aids.

Monique watches traffic through the passenger window, studying the faces of pilots they pass on the skyway.

Three

moving parts

.Rotterdam Boundary.AM/FK01—EU.2154.06.20.09:21

At freight checkpoint K on the outskirts of Paypal's twelve district, Ennis sits in the idling utility waiting his turn in line, peripherally aware of Amsterdam's outer walls through his windshield, but refusing to look at it.

The bomb is intended only as a diversion, there to hold Amsterdam's attention while he slips through its border unnoticed.

That's Plan A. Plan B is a little more convoluted. It relies heavily on improvising, some ad-libbing, and a lot of thinking on his feet. Despite the haste, Ennis has hedged his bets well. Either he gets in, or the nuke does. At least that's the way he sees it.

The PDA shows **2154.06.20.09:22**. In nine minutes he makes the call.

A pair of border agents wave forward the vehicle in front. It pulls into the inspection zone as agents scan it with specialized PDAs. Its occupants, a pair of couriers, get out and place their hands against the hood, letting the border guys search.

Ennis watches them get back in and drive onto the loading ramp, then the agents wave him forward. Pulling onto the marked platform, he gets out and throws a PDA on the hood.

Hanging strong is the damp smell of nearby rain. Dark gray skies threaten to let loose.

When the nearest agent sees Ennis he waves the other off. "Go talk to the next guy," says the agent, "he's flashing some indicators."

The other one walks back up the line while the first swiftly waves his own PDA over Ennis's. It confirms their transaction with the standard cha-ching.

"Nice. I assume this favor doesn't get anyone shot or maimed?"

"It's not that kind of package."

The agent laughs. "Who ever heard of a guy smuggling dope *into* Amsterdam?"

Ennis smiles back. "No one ever will, right?"

He gets back in and drives onto the concrete dock, backing up to a cargo arm.

A toggle on the dash splits open the ute's rear, which folds away to reveal a crate marked 'Medical Material—Do not open outside of Level 3 quarantine'.

Rain begins to drum down on it while Ennis gets behind the arm's control panel, the automated freight train sounding its five minute departure warning.

the bluff

Charging on its dashboard cradle, the PDA projects Amsterdam's Chief of Police on call.

"Chief Walter, my name is Ennis Pryce . . . I have a nuclear device within Amsterdam's borders. I intend to activate it if my demands aren't met by your government."

"This is a legitimate threat?" asks the Chief.

Ennis takes his eyes off the road long enough to look him in the eye. "I will legitimately blow you the fuck up if you don't take this seriously."

"Well . . . uh, what are your demands?"

"Check your q-mail," says Ennis, "I'll call you back."

The rented utility speeds through tight Rotterdam streets and laneways, tracing the border's outer wall.

the several demands of Ennis "crazy bomber" Pryce, in no particular order

2154.06.20.09:01

Dear Amsterdam, as you may or may not be aware, I am a disgruntled former resident of your fine city forbidden from returning under penalty of death. My crime was setting foot beyond your borders, just the one time.

Seeing as how I've got an A-Bomb now, I decided to dictate this list of demands. I really feel that these few changes could spruce the place up a fair bit for those fortunate enough to still live within the walls (and I mean 'fortunate' sarcastically, which I concede is difficult to convey via text).

I suppose I could be persuaded not to blow up Amsterdam if you get working on some of this stuff today, but you'll probably want to put all your guys on it, as I demand satisfaction.

That is to say, I order you to carry out the following demands, satisfactorily.

Abolish no leave/no return. I had to leave for an operation, and would have died waiting for clearance. It's just an all around dick move, makes you come off like assholes.

Open borders to cheap electronics and pahrmaceuticals. Not everyone can afford K-Racs. The Z-Rac is a capable alternative.

What's the deal with Geeves Stoole? I'm not saying I won't watch his show if it's on, but why does his sense of humor have to be so dry? From what I hear, big celebrities like him and the Zeniths are able to come and go as they please. Why are there restrictions on travel only for the rest of us?

More sporting teams, right? You probably get this from everyone, but in this case, it's coming from someone with the capacity to destroy your city if you don't. Just because you reject the outside world and its universalist ideals or whatevs, doesn't mean we can't see some athletes getting their armball on from time to time. I know there's no such sport as armball but everyone I know is so into football and I'm pretty sick of those guys.

How did you dudes fill up that moat around the city? Shipped in water, or was it all done with aqua ducts? Really curious about this. My friend swears it comes from an underground river but Dave can be totally belligerent when he drinks, and is usually only half-remembering the story he read.

I want to play some golf with your President. I don't really know how to play so he'll have to show me the golf moves and explain the rules, but he seems like the kind of guy I could get along with, despite the present circumstances, and if he doesn't play golf then that would be pretty weird for the president of a whole country don't you think? Wait, country? Or city? What are you guys exactly now? Province? *Principality*??

The bomb was aboard a flight that docked in Amsterdam yesterday. From there it was transported to a mysterious secret location. It may be in another airport, or possibly a museum. It is a mystery!

Okay so the bomb is somewhere in Amsterdam and is set to go off some time tonight or tomorrow. Or maybe even later in the week. Actually let's just say tonight or tomorrow.

Sincerely,

Ennis Pryce
"Crazy bomber"
11,284 Sky City Seven
PayPal 12 District
Rotterdam region—EU.

defcon

.Amsterdam Sovereign Economic Zone—EU.2154.06.20.09:35

"This isn't a serious list," says the President of Amsterdam, showing it to the other advisors in his briefing room.

Chief Walter—sitting opposite—shakes his head. "At this stage sir we're working without a psyche profile, Pryce may actually expect us to take the document seriously, despite the unfocused rambling."

"And how likely is it that he has a nuclear weapon?"

"They're not that hard to obtain outside the walls. It's getting savage out there in the cracks between their fucking hubris zones sir."

"Where are we on orbital detection?"

"We'll have three satellites tasked within minutes. If there's anything radioactive in Amsterdam they'll locate a signature."

President Gibbon takes a moment to process the potential outcomes, vacantly staring at the far wall. "This is going to sound a little cliché, but Amsterdam doesn't negotiate with terrorists."

"If the device is real sir, we'll find it. We know our city better than this guy. Once we locate his bargaining chip, negotiation becomes a moot point."

The President leans back in his leather chair, fidgeting.

"Clive, what do you think?"

The advisor by his side reads the list of demands. "Sir I believe the list is a diversion."

"How so? He wants our attention on the airports and museums obviously?"

"Sure . . . but not to stop us finding a nuclear weapon. This guy doesn't want to blow up Amsterdam. He wants to come home. The bomb—if there is a bomb—is just more misdirection. Pryce knows that a high-level internal threat will thin our ranks at the border. He's counting on it."

"Chief," says the President, "you should focus on airports and train terminals, and of course the museums. I want grunts there ready to search by hand if we get a sniff of it, but reinforce border checkpoints with a spread of TRG teams in case Pryce's notion is remigration."

"What should I tell him when he calls back?"

"Say we're working on his list—that it'll take several hours, but nothing he wants seems unreasonable. It doesn't count as negotiation if we're lying."

"I'll make it convincing—" says the Chief, interrupted by his ringing PDA. "Do you mind, sir? This may be him."

"Go ahead."

The Chief removes it from a pocket. "Answer. This is Walter."

"Chief I'm Jim Kraft in E-Warfare, I was asked to do a sweep of the city for some sort—"

"I'm up to speed, Jim. What did we find?"

On speaker, Kraft says, "There's definitely nuclear material on the island, sir. It's something pretty big."

Every man in the room maintains their stoic composure, an implied agreement.

"Where is it?" asks the Chief.

"We're having some difficulty with that."

"What's the problem?"

"The footprint we're seeing originates from within the island's central freight terminal. That's a seventy floor underground station—automated, with minimal access for maintenance."

The Chief registers Gibbon's paler shade, standing. "Keep me up to date." Pocketing his PDA, Walter says, "Sir, I'll let you know as soon as we have more. It won't take us long to find a way in."

The President nods, still mentally crunching the numbers.

As Chief Walter leaves the room, Gibbon turns to Clive the advisor.

"How certain are you that Pryce doesn't want us all dead?"

Clive sits on it a while. "Ninety-six per cent."

safety net

.Amsterdam Boundary Exclusion Zone.Outer Ring—EU.2154.06.20.09:46

At Checkpoint 31, border agent Luna Phranc is nervous.

The TRG guys who showed up twenty minutes ago won't say why they're here. Rumors are circulating that a terrorist threat was received. Now there's a utility truck speeding through slow point two, already halfway across the bridge.

Her orders are to shoot, but no one seems to want to be the first. Not even the tactical response team.

She turns to the nearest TRG guy and asks, "Do I shoot?"

Head shakes, eye glued to the scope.

The TRG all wait until the vehicle hits slow point three, then one by one they hoist their rifles.

threading the needle

When shots begin to ricochet off the hood and windshield, Ennis sinks lower behind the wheel.

Despite the utility's 'bullet resistant' selling point, he doesn't wish to quality test it here.

An RPG round curls out of the checkpoint's bunker, missing by a good ten feet, but close enough to give him a moment's hesitation. There wasn't supposed to be this kind of firepower at Checkpoint 31.

They figured him out, those sons of bitches. His final gamble, the one that's going to get him shot, if they fail to take him seriously enough. He shouldn't have carried on like a dickhead, trying to make them doubt his sanity. It should have been a focused, concise threat to their safety. Those fucks. They probably don't even believe he has a bomb. They should have had satellites overhead by now, they should have some understanding of the kind of shit they're in, by the time he's trying to gatecrash 31.

The explosion in his rearview barely registers. Ennis is focused only on the first of three concrete gates—slowly rolling closed.

Amidst a hail of small arms and tracer fire, his rental scrapes through with room to spare, already baring down on the second gate.

control

From inside the bunker Luna can clearly extrapolate from the sheer amount of ass he's hauling that gate three won't close in time.

She exits via steps at the back, looking around for a vehicle.

Running to the armored TRG troop transporter, she climbs into the driver's seat. A keycard hangs conspicuous from its sun visor.

home run

Sparks fly from the back left panel as it glances the second gate, whipping out the ute's rear wheels for just a second before traction control kicks in.

The third gate is only now beginning to close. It will be easier than it looked at the outset. Nudging 200 kilometers per hour, with open air between Ennis and Amsterdam, nothing can come between them now.

Nothing, as luck would have it, except an armor-plated TRG troop transport.

an armor-plated TRG troop transport

Luna throws the big transporter in reverse and backs it out, aiming for the third gate.

It rolls past the bunker and its adjoining wall, crashing into the far side, the sound of grinding metal and stone while it wedges in the opening, the closing barrier door crushing it slightly before seizing up.

Luna barely has time to brace for impact as the border jumper's truck rear-ends hers at high speed, triggering all thirty-six airbags.

kinetic sponge

Red crash foam bursts from the crumpling dash, slowing Ennis to a tolerable speed as he hits the windshield.

What's left of the utility shunts the armored truck forward and pinballs out onto the final stretch of bridge, coming to rest jack-knifed against the guide rail, interior cabin still relatively intact.

The foam hardens against shattered windows, dissolved quickly by its exposure to air, blowing away on the wind in fading chunks.

Ennis slowly regains consciousness while troops surround the crash scene, weapons ready.

Dazed, he shows them a remote detonator, flipping the safety panel open on its thumb switch.

"What is it?" asks one of the troops.

Another says, "Distress flare."

"It's a detonator!" screams Luna, climbing out of the transport's passenger side with her sidearm trained on Ennis.

Ennis kicks out the side window and crawls free, showing them all the remote, face numb, drooling, seeing spots.

Focusing the last of his will, he limps away from the wreck, again toward Amsterdam.

Luna follows close behind with a two-fisted grip on her pistol.

"Can't let you do that," she yells after him, trying to sound authoritative.

As Ennis turns back she stops advancing.

"You don't understand what's happening," he tells Luna. "I'm still in control."

"I get it—you wired yourself with explosives. But I don't think you wanna die today . . . so I'm calling your bluff."

He laughs, a little bemused.

She eases back the pistol's hammer. "I'll die defending this border. It's the oath we all took."

"Your oath . . . kills everyone in that city. Didn't they tell you what a serious fricking guy I am lady? I have a nuclear bomb. I'm not some jerk from Rotterdam who just wants to get in on your bullshit socialist paradise. I'm going to change things in this city. Tell people to stand up for themselves. Travel restrictions are just another type of despotism."

Luna's breathing begins to slow, unsure now. She shouts back over her shoulder, "Could he be telling the truth? Is that why they sent you?"

A couple of the soldiers behind her mutter to each other, sounding confused. Sensing the gravity more, she cautiously lowers the pistol, throwing a quick glance at the others, also lowering weapons.

Satisfied, Ennis staggers away, ignoring the dull throb of a broken leg.

Shock will get him the rest of the way. All he needs now is to remain conscious. When he's through the wall he will steal transport, and locate a doctor or veterinarian. Then he can hole up, pass out. Regroup. Plan the next three or four days. They will stop looking for him soon enough, move onto the next non-issue.

He doesn't see Luna drop to one knee and take careful aim at his centre mass. Unable to shake a slight nervous twitch, she hardens her resolve long enough to fire four times—two sets of two—catching Ennis between the shoulder blades. Each nudges him forward like a friendly shove.

He stops there, swaying, not falling.

Feeling the corners of his world fade to white, he looks down at pulpy black exit wounds. A crimson spatter decorates his favorite pair of sneakers.

Ennis turns away from Amsterdam, eyes meeting Luna as she advances on him, already reaching for the remote when he thumbs its switch.

Four

orientation

.Montana Peace Enforcement.LM/PE29—North America.2153.02.22.07:20

Imogen sits beside the Detective's desk, staring at a stain on the far wall. Detective Bentley, early fifties with a cop's gut and frayed tie, a man who exudes misplaced confidence and detached compassion, leans back from the feed display after putting the final touches on her statement, reading it over.

"I think we're just about done, Miss Harper."

With composure she asks, "Do you think you'll find this guy?"

He doesn't take his eyes off the display. "We sure hope so." But Bentley doesn't truly hope. He's seen a good deal of sanctioned wetwork, small and large. Given the victim's place within the management structure, he figures this one rates somewhere in between. Like any member of Enforcement, the detective knows that such investigations lead only to what they have coined 'the vacuum'.

He flashes a sad, almost reassuring smile and punches submit.

"Thanks for your time," he tells her, "I'll let you know if there's anything else we need."

phased

Imogen stands at a taxi rank outside Precinct 29 pulling her coat tighter, shuddering not from the cold itself but the awareness that snow has begun to fall.

Behind her a man exits the station obviously perturbed, kicking the front doors when they slide closed.

He yells "Eat a dick, cop faggots!" then flips them off through the glass.

The guy approaches Imogen, hands in pockets, seeming to calm as soon as the precinct is out of his line of sight.

"Lady?" he asks—softer than before—tapping her arm with the back of a gloved hand. "Are you waiting for a cab?"

She takes a moment to glance over. "Yeah."

"Where to?"

"Uptown," she says cautiously.

"Oh yeah? A cousin of mine lives up that way. Funny guy. Do you know the big fountain? His apartment is right over that. Steve. Know a Steve?"

Imogen shakes her head. "I don't—think so."

The stranger removes a cigarette pack from his coat, fumbling for one through leather gloves.

"Do you smoke?" He offers them to her.

"No . . . I quit a few years ago."

He lights it with a bright red Zippo. "So what are you in for?"

"Oh. No. I witnessed a murder. Sort of. I was sleeping."

"Ouch. Private or corporate?"

She shrugs. "What's the difference?"

"Private," he says, twisting his face to blow the smoke away from her, "Enforcement will go pick the guy up tonight. If it's corporate . . . then it basically never happened."

An officer walks outside and lights a smoke, looking at them briefly. "You people know there's a transport strike, right?"

Imogen sighs. "I got a cab here at lunch time. When did it start?"

"About twenty minutes ago."

She looks around—at the sky. "Do you think I can get a ride or something then?"

"I don't know," says the cop, headed for a diner across the street, "*can* you?"

The stranger tells her, "I guess we should walk."

Imogen begins trudging uptown, carefully stepping over puddles, followed by her new acquaintance.

"What was your name?"

"Imogen."

He offers his gloved hand, which she shakes with the tips of two fingers and a thumb.

The stranger says, "Otis."

"And what are *you* in for, Otis?"

"Well . . . kind of a long story. I imported some stuff that I shouldn't have."

"Drugs?"

"No, nothing like that. I paid a few fines, everything's cool."

"Okay." She hides a thin smile. "Why were you so angry?"

"They said I don't look a day over twenty. Kept calling me kid, Junior, even chuckles."

Despite the urge to pull away, she can't help but be mildly charmed by him.

As the snow begins to thicken he says, "Do you want to maybe get a drink? I know a place about three blocks from here. Maybe we can just wait out this strike."

velour

In the penthouse bar of a shareholder regional hotel, Otis and Imogen are able to supplement the pauses in conversation with long looks out at the view over swaths of skyscrapers and streaming air traffic, four hundred floors above the winding river that creates a canyon in the dense sprawl.

"Will you stay here?"

"Probably," he says between sips of scotch. "It's convenient, with the strike on."

"I thought you had to be a shareholder?"

"I am."

"In Lauchlap Michael? Why?"

"Why not?"

Imogen seems confused. "But then why not just live in one of those serviced apartments over in Two district?"

"I keep a small amount of stock in a lot of corps, it means free transport and accommodation in about two thousand territories. Probably seems like excess but I used to stay in three-star airport hotels every night, these cramped little concrete boxes that no one should ever have to spend more than an hour in."

She stirs a vodka and cranberry with its straw. "Employee housing is nice enough, but if I get those shares in my ten-year bonus I'll probably move. And of course stay here for at least a night. Even if it's just to say I did."

"How long have you been working for them?"

"Nearly . . . eight years."

He nods, finishing the scotch. "At college before that?"

"Yeah. Well, college then engineering school."

"What is it you do exactly?"

"It's a little embarrassing." She clears her throat and takes a sip through the straw.

"Go on."

"I help design starships," says Imogen, avoiding his eyes.

He hesitates. "You're kidding."

"Nope."

"Wow . . . you must be loaded, why don't you just *buy* some stock?"

Imogen turns suddenly cold. "I'm still paying off old debts."

"College loans?"

She stirs the drink again, staring into it.

"Something like that."

Otis waves to the waiter, holding up his empty glass. "Yeah, those can be a real cunt."

punch out

.Zibo Healthcare.MA/RD03—Asia.2167.09.02.16:39

Umberto swipes his PDA over a hardware sign-in station inside Regional Dispatch Three's rooftop hangar while Monique watches Medevac technicians rush diagnostic carts and fuel hoses to the returned helo, its next pilot already prepping his launch kit at one side. The pilot gives Umberto a courtesy nod, noting the stranger.

Dumping the helmet in a dented locker, Umberto throws a clean towel over one shoulder then looks at Monique, a bit unsure.

"So listen . . . I'm gonna take a shower, if you're going north I can give you a ride but otherwise you should probably call a cab."

"I can wait," she says, affable, taking a seat by the lockers.

drive thru

Umberto's ten year old Honda pulls into a Burger King car park by the highway onramp, waiting in line for drive thru.

"What do you want? It's on me."

Wipers streak across the windshield on intermittent.

"I'll just have whatever."

Looking straight ahead he asks, "Where am I dropping you after this?"

Monique thinks. "I don't know any place they won't tie me to. They're monitoring my friends and family."

"Hotel?"

"Not without access to my accounts."

"I don't want to seem insensitive," says Umberto, "but I can't exactly take you home to my housemate and ask can we keep her."

"You've already taken a lot of risks. Just drop me at a maglev junction if that's okay?"

Umberto hesitates. "You can crash on our couch for a few days—one week tops—but that's gonna have to be enough time for you to work on getting somewhere safe because me and Rolo just had this whole big thing about his girlfriend staying over, I won't bore you with the details but long story short, a week is the limit for all of our houseguests now."

She tries not to let him see the short moment of relief, but it's the elephant in the car.

"Thanks Umberto."

He drives up to the order box and lowers his window, leaning out.

"Hi, yeah, I'll take two whopper two-point-oh meals, a couple of shakes, and your curliest straws."

rolling plague

.Amsterdam Boundary Exclusion Zone.Outer Ring—EU.2154.06.20.09:59

A pair of MoD dropships touch down on the bridge just inside Checkpoint 31, engines idling as troops in radiation armor erupt from side doors, splitting up to check vitals on nine bodies spotted by satellite.

In eleven minutes the fallout will reach this side of the moat.

A pillar of ash casts its long shadow over the burning city. Despite the daily drills meant to ready them for precisely this scenario, it is the most surreal thing any of these hardened men have seen. Nothing could have been further from the realm of probability when they woke up today.

Kneeling over one of the bodies, blood pooling at his knees, a medic raises his arm, waving to the others.

the card

.Brussels Healthcare.OM/NV00—EU.2154.06.22.05:02

Ennis wakes to the steady tick of an ECG.

Sleeping in the chair beside his bed is a man he doesn't know.

The man sits up instantly when Ennis shifts, scratching the corner of one eye. "Mr. Pryce . . . I'm Lambert. I represent the Capital Expanse Alliance. But not officially. As far as anyone else is concerned, I was never here. I never sat beside your bed for three days waiting for you to wake up, because

I don't have a message from the Board of CFOs for you."

"Okay." Ennis looks around for water.

"*Officially,*" says Lambert, leaning forward, "I never told you that three days ago you accomplished what we've been unable to do in nearly half a century of mediation. And just to be very clear, I *didn't* offer you a well-paid position with Capital Expanse should you decide to further pursue a career in counterdiplomacy."

Ennis frowns.

Lambert stands, leaving a business card on the bedside table before walking to the door.

"Get in touch if you ever need anything." And with that, he's gone.

Ennis reaches for the card, wincing at undefined pain radiating through his torso.

A nurse hurries by his door, glancing at Ennis.

"Nurse?" She stops, backs up. "Can I have some water?"

The nurse simply glares, goes to walk away, but stops again. She spits on the floor between them.

waiting

The hospital's head of medicine escorts Ennis to the main doors, where a large crowd has gathered.

"This is about me?"

Dr. Trov nods. "The media tends to latch on to these things. There's a car waiting, over there."

"Who pays for that?"

"I don't know."

Trov waits there as Ennis exits through glass airlock doors, Enforcement officers in riot gear keeping the three-deep crowd at bay.

Ennis makes his way through the makeshift human aisle toward a waiting towncar—a nervous-looking driver by its open rear door.

One of the crowd begins to clap. Probably an attempt at being sardonic, rather than a congratulatory gesture, if he had to guess.

Then someone boos angrily in response, obviously missing the irony.

Another throws eggs that go sailing overhead, breaking on the street.

Then, rather abruptly, someone fires a shot.

Everyone in the crowd drops at the crack of gunfire, the cops all pulling their sidearms, patting themselves down for fresh wounds.

Half a block away, a hooded man bolts away at full tilt.

The Enforcement guys give chase while Ennis runs to the car, diving in back.

The driver slams the door closed with his head low, dashing around front.

It's now that Ennis notices the crowd gathering where one of the onlookers isn't getting back up.

As they pull away from the curb, he catches sight of the man giving CPR to a girl bleeding from her head.

equity

.ROTTERDAM TRANSIT.OM/UH45—EU.2153.01.27.08:26

The public ghettos between Paypal and Kruger districts are permanently darkened by MoD's overhead highway hub. No sound is present beyond the distant rumble of civilization. The once abandoned and recently reoccupied streets are a world unacknowledged on any official record of habitation.

Otis walks to the wooden shack's front door and rings the bell, throwing cautious glances around the pitch black neighborhood. Eventually, a display pops up at eye level. Keo can be seen leaning forward, eating what looks like canned ham.

"Come around back."

Otis goes to the side gate and squeezes through. In the middle of the back yard, protruding from the fake lawn, a metal bomb shelter door. He wipes the narrow Perspex window and peers in.

"Keo?"

Nothing. Otis twists the handle, heaving it open, smelling stale air as he climbs down the interior ladder.

"Keo." His voice reverberates in the tight space.

"Yeah bro?"

He walks into the living area between shelves of canned goods.

Keo lays on the only couch, finishing his ham, watching an infomentary on feed.

"Yo," grunts Keo, mouth full.

Otis slaps his feet off the couch and sits down.

"So are you in or what?"

Shrugging, Otis says, "I'll give it a shot."

Keo grins, spitting chunks of Spam. "My man!"

On the feed display, a window pops up with 'DOORBELL!' in big letters, live footage of the man buzzing it.

"Well hey, here's your big chance." says Keo.

"What?"

"I wanna see how you deal with this guy."

"Deal with who?"

"Some smalltime jerk looking for a nuke. I think he's into white power or something."

"I don't think so," says Otis, getting to his feet. "I wouldn't even know what to do."

"Just answer some questions and show him the hardware I laid out on the table. If he doesn't have the cash on him, act disinterested until he goes away. He'll come back if he really wants it."

Otis paces in front of the couch, studying the man at the door.

"Do you really think it's a good idea, selling shit like that to a Nazi?"

Keo laughs. "Ah, poor naïve Otis. Welcome to the *real* weapons industry, dude. If everyone who bought atomic bombs set them off don't you think you would have heard about one by now? I'm gonna go hide in the shower—but I'll be listening. Stay calm, you'll do good. Remember, this guy is beneath you. It's your world, he's just in visitor's parking."

While Keo shuffles off to the bathroom with his arms full of canned puddings, Otis turns to the screen. "Answer it."

"Hi," says the client. "We spoke on the phone."

Otis hesitates. "It's around back. Big fucking door in the lawn. You can't miss it."

layover

Reclining in his plush chair, Otis enjoys the last bite of the third-best steak he ever ate.

He feed-surfs a while, fidgeting, then waves at a passing flight attendant.

"Excuse me. Are we likely to arrive on time?"

She collects his plate and glass. "We're looking at about eight minutes holding outside North American airspace. Can I get you anything else?"

"Just a beer. Something imported."

The flight attendant smiles. "Sir we're over the Pacific—everything's imported."

"Got any space beer?"

"I'm pretty sure they don't make that."

Otis shrugs. "Surprise me then."

kings of kong

.Seattle Entertainment.MT/SC94—North America.2153.02.18.21:11

At a popular glass-floored cocktail bar on the top tier of Sky City Ninety-Four, Otis and the three gentlemen from Mungtao sit awkwardly on ergonomic glass bar stools around a glass table, each nursing an empty cocktail glass.

Otis looks down at his. "That last one seemed to have some Absynthe in it."

The reps all nod their agreement.

After a long silence the man opposite Otis—the eldest of the three—lifts his head to make eye contact. "The thing we mentioned. It's rather specific."

"I can do specific," he says without hesitating.

The man to his left—the youngest—says, "It's proven a tough get for several of your competitors."

"If I had some details . . . "

The first two look to the third. He checks each corner of the room before speaking. "We require a firm agreement with relevant NDAs be in place before divulging anything further."

Otis waves to the bar's concierge, who comes right over.

"Another round of drinks here please."

As the concierge leaves for the bar, Otis returns his attention to the clients.

"So at least tell me one thing . . . who am I getting involved with on this?"

"Again," says the third man, "that's not something we can reveal until we have your DNA on an NDA."

Otis thinks this over. "Okay. But I'll tell you this much, if you guys are getting me tied up with McPherson in any way, it'll be shitty for everyone at the table."

None of the men blink for a long time.

rectangles

.Zibo Residential.MA/M.EH44—Asia.2167.09.02.20:15

After ascending the lift, stepping into the nondescript eggshell corridor of employee housing complex forty-four in Medevac Asiatic's M District, Umberto leads Monique through the modest tenth floor apartment, tidying as he goes.

"Again, I'm . . . sorry about the mess."

"You should see mine."

He sidesteps into the kitchen, stacking dirty dishes, while Monique walks past into an adjoining living room.

She makes a beeline for the tall but typically narrow window, looking down at the city sprawl.

"I love your view."

"It gets old." Umberto places a pair of drinking glasses on the counter between them. "Water?"

"Sure."

"You can take a shower if you want, it's through the bedroom on the right—that's mine. Door on the left is Rolo's, he does sixteen hour shifts so best to just keep it down out here if the door's shut."

She turns to face him. "Do you have a bathtub in there?"

"Oh, yeah sure, I don't think I've ever used it but if you want to take a bath feel free."

"I just want to wash my feet."

Umberto stares a moment, somehow confounded by the idea. "I—okay.

I'll get you a towel."

"Thanks." She walks through his bedroom to the ensuite bathroom.

zen circle

When he walks into the bathroom—knocking softly on the open door—what he sees triggers a flood of emotion that feels somewhere between deep joy and overwhelming sadness, hitting him all at once.

Monique sits sideways on the small tub's edge with her back to Umberto, her dirty, faded blue jeans rolled up to the knees, feet soaking in ankle-deep warm water.

The harsh glow of fluorescent walls bring out waves in her dark brown hair, tracing its flow into the curve of her back, catching his eye, reminding Umberto of art that he can't remember anything else about, the association buried somewhere deep in the subconscious, frustratingly vague.

For just an instant he feels like his knees might give, resting a hand on the towel rack.

Like the memory of a dream, ambiguous and fleeting, he feels as if this scene has unfolded once before. Maybe another life.

She turns to him over one shoulder, smiling.

He hands her the towel then turns to leave, pausing at the door.

"I'll put some tea on."

Umberto steals another glance then leaves, confused by the experience.

diorama

After dinner, having insisted on washing up, she joins Umberto again in the lounge room.

He lays back on the couch's armrest, mashing a control pad. On the main feed display, an OLFBRPG plays out one of its strobe-light melee battles.

He scoots over as she sits down beside him, studying the game.

After sitting a while, she turns to Umberto. "Do you want to watch a movie or something?"

Umberto turns his face toward her without taking his eyes off the display.

"Oh, ah, yeah this is on feed . . . no pause button. If I bailed now, I'd really be letting the guys in my clan down."

"Clan?"

He affords her a quick glance.

"It's a gaming thing."

Monique looks around for something to read.

"How long does it take?"

Umberto shrugs, thumb and trigger finger still twitching violently at the handgrip.

"Maybe an hour?"

She asks, "How much of this stuff do you play?"

"Probably more than average."

Umberto jumps off the couch, punching the display.

"Yes! Eat it you fucker!"

He flops back down.

"Is it over? Did you win?"

"Oh I won . . . two and a half experience points."

"Is that good?"

"It isn't bad."

She clicks through the pages of a local feed guide on his PDA.

"Now watch me jump on the back of this wolf and ride to the sea."

"Do you mind if I lay down in your room?"

"Sure."

She leaves quietly.

glimmer

.Worldsplus Accommodation.WP/SH14—EU.2155.03.02.21:13

Ennis stands on the balcony of his hotel room, smoking a blend of cannabis and coffee rolled into a cigar skin.

A billboard at eye level degausses when an old cargo ship roars by, catching his gaze.

He watches the light hauler ascend at full thrust until it breaks orbit, dissolving quickly to become another speck of light amongst a bedazzled night sky.

Months have passed here, living free and anonymous on Worldsnet's dime. It's well past time to leave what is, essentially, another kind of prison.

He's not looking for Amsterdam, or a place like it, anymore. He doesn't like to remind himself of anything associated with that part of his former life.

Mars always interested him. It's civilized, corporatised, but with a leftist twist that makes it seem less foreboding.

If he lived on Mars, he could probably get a dog. He never had a dog before.

FIVE

tornados are touching down on north america

.BOISE MILITARY.DEPARTITION—NORTH AMERICA.2153.02.19.08:02

Amidst decaying navy proving grounds in the iTunes desert, the lone Bosworth *Echo*, a miniscule, twin engine pleasurecraft not produced since 2141, approaches low and silent despite the lack of any military activity here these past ten or so years.

It puts down on charred concrete tarmac once used to launch Truman class naval repair ships, dwarfed by what remains of empty tankers and toothless gunships. Even the tugs tower over it.

The base and surrounding infrastructure sits empty amongst a landscape seldom visited. Rot is in everything, at a cellular level. The skeletons of not only ships and buildings, but animals who came here with the same assumption of permanence as the men who would later return this stretch of land to the Earth.

Most assume that the tract of desert is still restricted under MoD partitioning, but navy pulled out of Boise more than a decade ago. Now it's another unmapped territory, open to exploitation should it ever become commercially viable.

Wind sweeps dust across the sprawling flatland in sheets, sandblasting the collection of admin and infirmary buildings to the north. The remains of a long haul supply ship rusts half-inside the closest hangar.

Otis pulls the jacket's hood down over his eyes stepping out of the Bosworth, walking at a brisk lean toward the buildings, squalls thrashing at him.

A weather alert vibrates his PDA, but Otis knows all about the incoming storm system.

On the horizon lightning burns under black cloud, while forming twisters loom like crooked fingers.

behemoth's footprint

Otis reclines in the dusty chair of a mid-level data clerk's cubicle, juggling several displays, each scanning long lines of system and network code.

In his pocket, the PDA vibrates again, more urgent than before.

Still reading, he mutters, "Audio only."

"Otis?" The female caller's voice wavers, degraded by approaching lightning.

He blinks, looking away from the displays.

"Hey. What's happening?"

"Am I going to see you this week?"

Otis hesitates. "There's a job on right now . . . I'm not sure how long it'll take."

A pause, then she asks, "Where are you?"

On one display, the scrolling code has stopped, highlighting several outlying strings.

Otis leans over the keyboard made of light, forwarding the log to his PDA.

"Deni I'm right in the middle of something, can I call you back?"

Fading in and out, she says, "Probably."

small writing

.Santiago Residential.GH/EH56—South America.2153.02.20.18:30

In an apartment overlooking Gallagher's upper One district, Otis watches Keo mix a second Tom Collins for them both.

He asks, "Hey what have you heard about a ship called Behemoth?"

Keo pauses, frowning purposefully at Otis. "Don't tell me someone has you chasing Behemoth on the side."

Otis shrugs. "Even if I was involved in that kind of thing, I imagine it's not something I'd be allowed to discuss."

Slowly shaking his head, Keo sighs. "Bro . . . we've all considered it when things are lean."

"Who has?"

"Everyone. Some guys tried and failed. Others were smart enough to never go down that road. McPherson has been pitching it at every new guy for a good fifteen years."

"It wasn't McPherson who hired me," says Otis, "I specifically told the guys I wouldn't get involved with anyone outsourcing for them."

Handing him the drink, Keo says, "Yeah? Look into it. I can pretty much guarantee you're dealing with McPherson on the financial end."

"Any particular reason?"

"Experience."

Otis sips his Tom Collins, habitually scanning the markings on traffic outside.

obligatory roses

.Santiago Residential.UTM/EH1404—South America.2153.02.21.07:12

When Otis steps out of the shower Deni is brushing her teeth at the sink. He moves around her to the bathmat, awkward in the narrow space. The lower tier apartment in the shadow of Sky City Sixteen, deep within Ung Tau Maria's central residential infrastructure, wasn't built with things like employee comfort, or dual occupancy, in mind. Drying himself with detached automation, he mentally rehearses the string of flights he will need to catch today, avoiding her eyes in the mirror.

She smiles at Otis when he finally glances up, a smile he returns with only his mouth.

Spitting toothpaste, she says, "Keo called. I said you'd call back."

"Okay."

"Do you want to get breakfast?"

Otis turns away, pulling on a dress shirt. "I can't. There's a brunch in Asia."

Deni stays a moment, studying herself through the fogged mirror.

laylines

Nine clicks above the Pfizer-Pacific Ocean, aboard the 08.11 shuttle to Tokyo, Keo hands his tray to the flight attendant with a smile, folding away the table before reclining. He reaches across the empty seat beside him, touching Otis on the elbow.

Otis removes earbuds, pocketing them.

"Okay here it is," says Keo, "I offer you this in the hope that you'll stop drinking alone for just tonight."

Otis sighs. "I don't feel comfortable in strip clubs."

"About three years ago one of my guys exploited an MoD relay, he had access to six months of their secure feed before they shut him down."

"How did he—"

"I couldn't tell you if I knew. It's not the point. So amongst the resulting data we logged there were several production-level blueprints flagged as 'unused'."

"Okay . . . "

"One of the larger ship designs was a twelve-click unit codenamed 'Mobius'. Its geometry made it seem a prime candidate for rejection, given that it was lumped in with freighters and transports. To us, it looked more like an orbital vessel. Too small of course, but quite obviously laid-out with ground support missions in mind."

Now Keo has his full attention.

"A Lauchlap Michael analyst freelances for me on naval things, I asked him to look over it . . . he called me a minute after receiving the blueprint and pointed out that 'Mobius' has only ten centimeters of headroom on every deck."

Otis frowns. "Huh?"

"It's scaled down. By his guess, a hundred to one. The Lauchlap guy couldn't think of any practical reason to modify a design like that other than maybe pushing it through a committee unnoticed. He was surprised to learn that the blueprints were production level, because MoD makes its cuts during the first few drafts."

For a long time, Otis thinks. Then, "Wait. It's a twelve-hundred kilometer-long OV?"

"Hidden amongst a package of ship designs that shouldn't exist. Broadcast as priority, on secure feed."

"Even if someone could build it . . . "

Keo palm-shrugs. "It would be grossly oversized, unsuitable for service on any habitable planet."

Otis seems uneasy with this.

"So here's the lead I had, that none of those other guys did. The engineer felt like he'd seen the rough design somewhere before. Besides contracting for me, he hasn't spent a whole lot of time outside Lauchlap Michael."

"Did you ever chase that up?"

"I started following a few employees, some names he gave me. For about a week, as it bored the shit out of me. None of them were entering secure areas."

Otis says, "A project like that would be compartmentalized."

"Right. It would be very hard to find an initial in, and even then, they might not know jack."

pay streak

.Montana Residential.LM/EH19—North America.2153.02.23.05:58

The first thing Otis is aware of as he wakes, before the woman or the unfamiliar bedroom, is the lack of a hangover.

For an instant he watches her, piecing together the sequence that brought him here. He swings his legs off the bed carefully, his eyes on her as he crosses the room to where she hung the jacket.

From its pocket he takes her PDA, placing it on the desk as he sits down at the small feed station.

"Feed," he tells it softly, "Lauchlap Michael PN."

As the window flickers to life, Otis throws another glance in her direction, the sleeping woman bathed now in its blue glow.

On display, the Lauchlap Michael private network prompts him for voice login.

Otis whispers, "Login manually."

At his chest, a lightboard appears. He types, 'Imogen Harper', then her twenty-four digit passcode.

tree/green

.Oporto Residential.AT/SH08—EU.2159.09.15.12:39

“At word association,” Sylber announces out of nowhere, “everybody cheats.”

On the rooftop café within The Amsterdam Trust’s small residential subdivision, Apple continues to read. It’s the kind of stupid shit he’s constantly coming out with, sitting there all proud of himself, expecting that she’ll offer some sort of considered response.

When his eager glare becomes unbearable to her she offers, “How’s that?”

On cue, “Some people take an instant to process it. Some take seconds. They all cheat. The hows and whys, the roadmap you lay out with every word . . . you telegraph indicators of the person you want to be, not who you are.”

“Okay.”

“In theory the exercise gives you a direct line to the subconscious, but the conscious mind is so good at rerouting it that we fool ourselves into thinking the word association is proof of who we hope we are.”

Now Apple looks at him across the table, on the edge of annoyed. “So you’re saying it’s not accurate.”

“No, it’s accurate,” says Sylber, “but not in the way it was designed.”

“I don’t . . . know what you mean.” She sips red tea, taking note of the dying afternoon sun.

“The people who answer right away, they’re the best liars. They cheat better at everything because it’s just the way their minds work.”

Apple fights the urge to roll her eyes. “What about honest people? Wouldn’t they also answer right away?"

He smiles, shaking his head. "No one is exempt from the rule. It's universally applicable."

"Why?"

"There are no honest people. Everyone lies about something. Especially about themselves. Even if it's only the small things."

She places her empty mug on the table, watching Sylber's pupils.

"So what do you lie about, Sylber?"

He stirs his Coke before meeting Apple's eyes. "Everything."

autumn sun

On the ride home through their small slice of corporate territory, Apple watches scenery pass from the passenger side of Sylber's Mercedes.

"It's getting colder," he says, reaching for conversation. "They said on Vanderbolt that continental weather services are starting to break down. No one really knows why, the tech is all sound . . . it's just the weather that changed. It's less malleable now, or something."

Apple just doesn't respond. It's too vapid, even for Sylber. As they turn for the garage beneath her apartment complex she taps a shortcut on her PDA, steel doors rolling open up ahead.

Sylber pulls up to her elevator, checking the time. "Well . . . see you when I see you."

She exits the car, watching Sylber Giles drive away.

soften

.Oporto Commercial.AT/CBD01—EU.2159.09.26.10:18

Apple sits by the window of her 540th floor office suite in the Amsterdam Trust's central business district, watching traffic pass, her faraway stare belying darker thoughts to Sylber as he enters with a knock.

He tells her, "We need you on the roof. Wheels up in ten."

Snapping to, she looks at Sylber. "Are we sanctioned or what?"

"Elysium's still processing the request. We'll know by tonight."

Apple gathers her few accessories from the desk then follows Sylber through the bullpen of analysts working hard at looking slightly busier as they approach.

ubermodern

.Ombudsman Transit.OM/HSBW12—Shipping Space.2159.09.26.14:33

Aboard a private shuttle traversing the hyperspace biway between Earth and Mars sovereign space, in back, amongst the cargo nets and crew kitchen, Sylber speaks quietly into the PDA he holds to his ear.

"What kind of timeframe would I be looking at, if there was no embargo?"

The woman on the phone says, "Three days. It's not ideal, but it's where we are."

Sylber ponders this for a time, obviously uneasy about it. "Yeah, okay . . . lock it in for Monday. We'll talk when I'm done on Mars. End call."

He re-enters the main cabin and sits back down beside Apple, who doesn't look up from her novel.

"Was that Elysium?"

Sylber watches one of the flight attendants, pouring coffee for the group of mercenary lobbyists hired by Apple.

"No," he says, "that one was personal."

semi-circles

.ISIDIS ACCOMMODATION.WF/LH03—MARS.2159.09.29.08:07

Sylber Giles enters Hotel Complex One in the Waldorf accommodation sector by its side entrance, making a beeline for the rear conference room, descending stairs to the row where Apple sits watching the guest speaker.

He leans over her, whispering, "We're sanctioned."

Apple smiles while Sylber leaves.

errant

When Apple returns to the company suite, Sylber and two of the lobbyists are at three corners of the common room, each speaking low and urgent into their PDAs.

Sylber ends his call and turns to Apple, unsure what to tell her.

"There's been an incident," says Sylber, "some things I was working on have gone south."

"What things? What did you do?"

He asks the lobbyists, "Could you give us a moment?"

Still on their calls, they file into the hallway murmuring.

"I made a play on the Pryce thing behind your back to give you plausible deniability if things went bad. They went . . . really bad."

"What play? You just said we were sanctioned. Why would you even—"

"I set this in motion when it wasn't looking like we'd get permission to extract him." Sylber looks at his feet. "I outsourced the hit to a Mars based retirement firm."

"Hit? You tried to have him executed? By who?"

He doesn't respond.

Apple says, "Oh Sylber . . . I know you didn't just entangle us with an unsanctioned company."

"Not in any way that can come back on us."

She sighs. "How are you this stupid?"

"I had limited time to make the call, so I went with my gut."

"Jesus . . . "

"I got it wrong. I'm sorry."

Apple walks to the tall window, her back to Sylber, looking at nothing in particular. "That's not the way this works. Plausible deniability is always off the table."

"It won't happen again. I'm sorry."

"What exactly did happen?"

"They put eyes on the ranch for days leading up to the attack, and had him sleeping between about two and six. It was supposed to be quiet."

Apple slowly shakes her head. "The one thing I wouldn't assume about the guy who killed a million people is that he gets regular sleep."

"It seems that was the case. He shot three of the four guys who went in there."

"And the fourth? Can we trust this guy?"

Sylber slowly shakes his head.

snakes and ladders

.Montana Operations.LM/SD344—North America.2153.04.10.15:43

Imogen sits at her desk in Nav Design on the 312th floor of Blackbird Tower, watching the PDA on it, listening to hold music. She counts the thirteenth minute on its clock display.

When someone finally does pick up, it's a different man than before.

"Imogen?"

"Who am I speaking to?"

"That's not important. You only need to know that I work here in head office."

"I—okay . . . are *you* able to tell me what's happening with Longhorn?"

A silence at the other end. Then, "Imogen I need to ask you to take a step back from that for now."

"I'm sorry?"

"We're investigating the issue. It's with our best engineers."

"Oh! Well that's great, I'm sorry but I got the impression that no one was really taking it seriously up there. I didn't mean to cause so much fuss, it's just, you know. Ticking time bomb and all." Imogen taps through file directories on her display. "Are you able to put me in touch with your lead? I've logged about seventy hours of simulation data relevant to the fault."

"You can send it to our general q-mail, and we'll pass it on."

Imogen pauses, reassessing, without wanting him to hear it in her voice. "Can I ask why I'm being pulled from the BD-77 fault?"

"You have bigger fish to fry. We need your expertise on something down here . . . I don't have all the details yet but you'll know what's going on by tomorrow night. Expect a transfer order and job description in your q-mail before then."

"Am . . . I being demoted?"

"Quite the opposite," says the man at head office. "Your paygrade was recently bumped, with sizeable shares. I've probably already said too much . . . no doubt your line manager will want to share the news. Keep an eye out for the confirmation, okay?"

For the longest time, she doesn't know what to feel.

exodus

Monique is awoken by strong hands, wrapped around both her wrists. They shake her twice, sharp.

"What are you doing?"

Leaning over her, Umberto whispers, "Do you trust me?"

After a moment she nods.

He pulls her out of bed, through the living room, the front door already open.

Half-asleep, she asks, "Where are we going?"

He drags her into the waiting elevator and punches '255' on the display.

bird of prey

Umberto pulls Monique sprinting across the 255th floor skywalk connecting Tower C with B, her feet barely touching the carpet.

Through the glass ceiling, he's aware of the banking jumpship on its landing approach to C. It makes no sound.

At the entrance to Tower B, Umberto swipes his PDA, the glass doors sliding open.

He drags her inside, Monique seeing the jumpship too before they're descending stairs two at a time.

chinatown

From a 243rd floor hallway window in Tower B, Umberto and Monique watch flashbangs light up his apartment.

Gas wafts from the bathroom's small extractor vent. The narrow windows all fogged.

Umberto takes her hand again, guiding Monique back to the elevators.

When they're inside he selects the basement light rail terminal, feeling his stomach lurch when they move.

She asks, "What do we do?"

"I can get us out of Asia," says Umberto, "but then we should talk. I think this is a good example of the need for specifics on my part."

junket

.Elysium Feedcast.CNN/FS811J—Elysium.2159.06.20.18:28

Ennis and Kirk Vanderbolt sit on the well-lit stage behind Vanderbolt's V-shaped desk, listening to the man beside them.

Facebook Senator Giancarlo Maxwell says, "When the first nuclear weapons were detonated on a human population it sparked an arms race amongst nations so fierce that it took all of humanity's diplomatic weight to halt the snowball effect of proliferation. With Amsterdam you've created a similar race, only this time amongst individual terrorist factions. Now posing a legitimate nuclear threat is in vogue for every outfit from animal rights to Islamiyan offshoots. In the age of dispersed infrastructure, splinter corporatism and flea market atomic bombs, there are no nations left, and no diplomats to stem the second tide of propagation."

Vanderbolt spins in his chair to face Ennis. "Mr. Pryce?"

"He has a point," says Ennis. "I mean he's reading it from palm cards, sure, but a salient argument nonetheless."

Vanderbolt pivots back to Maxwell, "Do you believe that we'll see some sort of . . . I don't know, *copycat* attacks in the near future, Senator?"

The Senator offers a quick shake of his head. "I don't think it's wise to speculate on something like that. I wouldn't rule it out though. Probably. Yes, absolutely probably."

Looking into A camera, then Ennis, Vanderbolt says, "We're running short on time, is there anything you'd like to add before we wrap up?"

Ennis shrugs.

"Well, then that about does it for it for this Amsterdam day, Thursday June twenty, twenty-one fifty-nine. Mike Lewis is—"

"I thought this would be cathartic somehow," Ennis cuts in. "When you had me on here for the one year anniversary I just wanted to own up, let people see me and know what I did. Give them a face, to put with the anger. When your producer approached me for the five year 'Frank discussion about Amsterdam', I wasn't under any kind of illusion, Kirk. I knew they'd sit me down here and trot out a leftist with proliferation or zealot with the corruption of morality."

Kirk clears his throat, taking a sip of water while Ennis looks up at the studio audience, really making the effort now to tune out the lights and guage the look on all those faces he avoided when coming out here onstage. He gets it. Sees how much spite there is still.

"I don't know why, but I thought it would help somehow. I keep telling myself that if I just give everyone a place to unload their vitriol, maybe all the angst will be spent and the world will move on. Then, who knows, I could move on too. But it's," he laughs a bit, coming unhinged, regaining, "it'll never be like that."

Again, Kirk Vanderbolt is unable to put together an impromptu segue. He looks at the Senator hoping for a response, but the older man seems to be physically sizing up Ennis, clearly disgusted and spoiling for fisticuffs. A rising murmur amongst the crowd, while even the crew seem to lean around their equipment to get a better look at Ennis Pryce.

Someone in the audience shouts, "Maybe you should kill yourself."

"Hey!" snaps Vanderbolt, pointing at the guy who said it. "What did I say before we went to air? Can we get him out of here please?"

Ennis wears a defeated, affable smile. "It's okay, Kirk. It's probably the most common reaction people have. And it's more than fair."

A commotion at the back of the room as the man is briefly led away by security.

"I won't be back for the ten year, to be honest. I've heard it all before, and I'm done. To the people who lost their loved ones . . . again, I am so very sorry. I can never make things right. To everyone else—if you see me on the street, please know in advance, I probably don't want to talk about it."

Ennis stands, unclipping the lapel mic.

To camera, Vanderbolt says, "Mike Lewis is up next with Sol's Failiest Fails. Goodnight everyone."

park life

.Elysium Unpartitioned.OM/UP22001—Mars.2159.09.28.13:02

On a three thousand square click subdivision bordering the Pepsi Cola desert, a lone ranch house sits amongst rolling foothills, overlooking the epic endless flats of Mars's largest desert plane.

Ennis, bearded now with long, tangled hair, exits the house, hauling a weighty bag of dog food, setting it down beside a dented feed tray in the front yard. He tears open the bag before emptying its contents into the trough, kicking the thing with his steel-toed boot, returning to the front door as a pair of cows come galloping around the house, racing for the feed.

island in the sun

"I don't think you're hearing me," says Ennis, the display tilting to follow him around the room while he paces, "I'm not looking to be *anyone's* employee. It's not personal. It doesn't relate specifically to you guys. I'm just out of that whole loop."

At the other end of the conference call, Willard Norris says, "We're not offering you a job, Ennis. We're offering sponsorship."

Ennis stops, looking at the display. At Norris, sitting head of the Worldsnet Alliance boardroom table.

"What's the difference?"

"Sponsorship is free range. You don't answer directly to any superior, not even the board. Alliance objectives are dispatched daily via an internal ticker . . . you take the ones you wish, leave what you don't like. You're afforded a generous operating budget reviewed quarterly by the board, allowed to allocate resources as you see fit. Our equipment and personnel would be at your disposal. If, after a year, either party is unhappy with the arrangement, we go our separate ways."

"It sounds too easy."

"Trust me, it isn't easy. But when you get something done, as the man in charge? It's satisfying."

"I'm plenty satisfied," says Ennis. "I like it here."

"Then I'm sorry to have wasted your time."

On display, 'Call ended by caller.'

other people's problem

Ennis is woken by the klaxon siren of a flashing alert above his bed.

It reads, 'Perimeter Alarm!'

"Attention," says a digitized voice, "the north-western border of your safe zone has been encroached by potentially hostile forces."

Already pulling on pants, Ennis runs to the bedroom window, peering out across the cold, foggy grounds.

"How many?"

"Four," it says, "on foot. Armed with assault rifles, carrying composition explosives. Do you require assistance by local Peace Enforcement at this time?"

Ennis doesn't reply.

"Do you require—"

"No," says Ennis.

He enters the walk-in closet, punching digits on a floor-to-ceiling safe. From it, Ennis takes a GA-12 large bore assault rifle and his holstered Sig sidearm.

A pair of holographic lenses flicker on above the rifle when Ennis lines it up on the window, getting used to the feel of its stock against his shoulder again.

The security system asks him, "Do you wish to dispatch drones?"

And Ennis says, "No, not yet."

Ping

Keeping his head low as another volley of tracer fire streams in, Ennis reloads the GA-12.

The rounds pound down on the rear of his overturned refrigerator, denting it without penetrating.

Laid across the open back doors the sturdy alloy fridge provides cover as Ennis swings the assault rifle over it once more to fire off a clip in three round bursts at the man sheltered by his tool shed.

Chamber clicking empty, the sight picks out his attacker's shape on infrared, falling face-first. Some luck finally.

Ennis makes a break for the main hallway, taking spiral stairs three at a time to the ranch home's second floor. Crossing the billiard room, he smashes out its window with the rifle, aiming at the guy crouched behind the front yard feed trough. Spotting Ennis's movement, he fires first, peppering the window frame and wall, forcing him back in. Downstairs, Ennis hears the sound of breaking glass.

"Attention," says the security system, "the kitchen window has been compromised within your safe zone. Do you wish to dispatch drones?"

"No," says Ennis, firing back at the man behind the trough.

He runs from the window as more gunfire punches through the outer wall, falling back to the spiral staircase.

Here he slows, creeping toward the top step with caution, line of sight giving him the half-second advantage over a masked gunman ascending with matched caution.

Ennis lets his trigger finger hover for just an instant, steadying his breath, lining up on the balaclava between desperate light blue eyes.

diminutive

Nestled amongst a natural rock formation in hills overlooking the house, Hallen Wilde receives another pulse of garbled static on the earpiece.

This one sounds more urgent than the last.

His starlight scope trained on the house's side entrance, he pans it quickly to the rear window, then back.

"Clyde," says Hallen into the mic, "I'm not reading Jorge or Damien."

Nothing.

"Clyde?"

Again, he moves the scope to the rear window, startled by the human shape looking back across its own rifle.

Hallen rolls aside, behind cover of the rocks, as large caliber ammo comes at him in threes.

Breathing red dust, keeping low, he shoulders the rifle and rolls up his kit, retreating through the hills via a well-worn hiking trail.

transcript

"My name's Ennis Pryce . . . I live on the big property out at the edge of your zone . . . I'm not even sure if you guys do my area."

"Okay Mr. Pryce I have your details in front of me, what's your emergency?"

"Some guys came tonight . . . invaded my ranch, attacked me with . . . guns. Grenades. Two of them had C4 I think. Their PDAs are locked but I have them here and I guess . . . your people can probably figure out if . . . I think the plan was to blow up my house."

"I'm dispatching units to your address Ennis, please stay on the line with me and we'll figure out what you need to do when they get there. Where are your attackers now?"

"I have security logs. IR data and footage. I'll provide all of it."

"Ennis have the men who attacked you fled the scene?"

"One of them. There were four."

"Where are the others? Can you still see them?"

"I shot them with my rifle. It's licensed, you can check."

"We will, Ennis. But what I need for you to do right now is place any weapons you're carrying in a well-lit area and keep your hands where our officers can see them as they arrive."

"I can do that . . . I won't cause you any trouble. The logs will verify . . . self-defense."

"No one will say otherwise until an investigation is in place. For now I must advise you that anything you say here is admissible as evidence in future proceedings."

"They attacked me in my home. The logs will clear me. I just . . . I've had some trouble in the past. I don't want you guys to think that I'm going psycho out here because of what happened on Earth."

"Ennis I'm not familiar with your situation but a unit is thirty seconds away from you now, you'll be able to tell the officer on scene everything you've just told me."

"It wasn't me this time."

"Take long, slow breaths. Calm down. Take a seat, and remember to keep your hands empty and clearly visible at all times."

"I appreciate your help. What was your name, sorry?"

"I'm not allowed to give you that, Ennis."

"That's okay. Thank you anyway."

"They're landing now."

sunrise

.ELYSIUM PEACE ENFORCEMENT.PI/PE49—MARS.2159.09.29.08:12

Ennis exits the small precinct building with hands in pockets, head down. Scanning the courtyard and car park with a quick glance, he takes the PDA from his coat pocket, telling it, "Call a taxi."

From the car park, a well-dressed man approaches him. "Ennis?"

Ennis stiffens, taking a step back, looking around for an exit strategy.

"I have a towncar. I can take you."

Ennis blinks.

"I work for the Worldsnet alliance. I'm Jeffe."

"Take me where?" asks Ennis.

"Wherever you need to go. We heard what happened."

"With the soldiers?"

"Yeah."

To Ennis, the guy seems genuine, but anxiety remains.

"Listen," says Ennis, "I have a gun. I know you're probably not about anything sinister but I have to be careful now. If you're fucking with me, if this is some sort of trick, my last goal in life will be the putting of a bullet in your head by me."

Jeffe slowly nods.

"That sounds fair, Ennis. They let you carry a gun in the precinct?"

Following Jeffe, Ennis says, "You can call me Mr. Pryce."

steak

With the towncar idling in the front drive, Ennis and Jeffe walk to the main house.

"What's that?" asks Jeffe, looking off to the lower ground by the house's rear.

"Hm?" Ennis follows his eyes.

"Did they leave one of the bodies?"

Ennis stares for a while before running down there. Jeffe follows at a jog.

When Jeffe catches up, Ennis kneels by the head of his deceased cow, petting it, sobbing.

"Is this your cow?"

Ennis nods.

"They shot it?"

"I don't think so," Ennis manages through the tears. "Maybe poison."

"Do you have any others? Maybe you should check them."

Ennis freezes, suddenly standing, looking around the property.

"Trevor!" he shouts, clapping. "C'mere girl!"

He makes for the foothills, starting to cry again, shaking it off.

Using the high ground, he scans the surrounding desert.

"Trevor! C'mon!" His voice cracks.

Now he spots the feed trough, dashing to it, kicking it hard.

Ennis waits.

"Trev!"

Nothing.

Jeffe joins him again here, shrugging.

"Maybe they ran off."

"Trevor doesn't just run off. They could have taken her."

"So . . . you only have two cows?"

Distracted, Ennis kicks the trough again.

"Yeah."

Jeffe looks at the empty Dog Chow bag.

"And you feed them dog food?"

"Yeah."

His next words, Jeffe seems to consider carefully.

"You know that could be it."

"Could what?"

"They put meat in dog food . . . cows don't eat meat."

Ennis frowns at Jeffe.

"Trevor and Graeme eat meat all the time. It's all they eat. You're probably thinking of horses."

"I'm pretty sure it's cows too. They eat grass and stuff."

"There is no grass. It's the desert."

Uncomfortable, Jeffe says, "I hate to say it . . . "

"Fuck off," says Ennis, then the tears are back. "What the fuck, man. That's some bullshit. You're out of your depth okay?"

"I'm sorry."

"If she's sick, she might have wandered into the desert. Get the car."

Jeffe runs back to the car, bringing it up to the house.

retribute

Driving their all terrain vehicle across the Pepsi Cola desert at four hundred sixty-seven miles per hour, Jeffe turns back to Ennis, who still watches the horizon through his window.

"There are trackers living here in the region," says Jeffe, "we employ freelancers. I can make a call."

"Okay."

"So is this all about Amsterdam, do you think? The thing from earlier?"

"Yeah. It's Amsterdam. Everything's about Amsterdam."

"Must be hard."

Ennis asks, "Can your trackers work at night?"

"Of course." Jeffe waves a hand over the center console. "Feed, jazz audio, something bluesy."

As soft saxophone fills the car Ennis leans back in his seat, eyes heavy.

"Do you want me to drop you back at the house?"

"I can't be there tonight . . . you better take me to a hotel."

Jeffe turns the wheel, bringing them around, back through their own dust.

"A good hotel?"

"That depends," says Ennis, "on who's paying."

"The alliance would be happy to cover your costs."

"This doesn't mean we're engaged or anything, Jeffe."

"You're breaking my heart Mr. Pryce."

Six

failsafe

.Comodoro Riviera Commercial.KVK/CBD01—South America.2153.07.27.17:59

With his feet up on Keo's desk, Otis fades in and out, fighting sleep while the five displays sift MoD's pilfered documents for keywords and images.

Outside, on the wind-swept smoking deck of Keovirak Systems' head office in its block-wide central business district, Keo snoozes on lawn furniture, drunk.

When Otis's PDA vibrates on the desk beside him, he ignores it a moment before glancing at caller ID.

"Audio. Hey, how's things?"

On speaker, Imogen says, "I'm good, yeah. What's been going on?"

"Same olds, really. Work. Running around all over the planet. Hotels, couches, other people's toilets. You know."

A pause, then she asks, "Well I don't know if you're planning to be in North America this week but I have some flex time coming up and no real ideas on how to use it."

"Go on . . . "

"Maybe we can check out a movie or something."

Otis watches Keo through the window, slowly nodding.

"Listen I've got some seriously fucking pointless data crunching going on, it's draining my will to live right now but it's got to be done, and it's only available to me here at my friend's office in Comodoro."

"Oh, well that's cool. Another time maybe."

Otis stands and turns away from the displays, away from Keo.

"Listen . . . why don't you come down here for a few days? I have miles. More than I could ever use."

"Hmmm . . . " She draws it out, like she hasn't decided. "Yeah okay, that works too."

35907609348760987.6905.class

Aeronautics internal memo B.12.97630: Orbital Vessel Failsafe Platforms
ATS CLASSIFICATION 8080

For every planetary orbital vessel (heretofore referred to as OV) there is a failsafe.

The failsafe rig is a control and communications platform located within restricted deep space, a station constantly linked with its designated planetary OV, manned by approximately four hundred crew, and protected by a specialized naval fleet and SEAL platoon.

The failsafe platform is considered as valuable an asset as the OV itself, given that the failsafe's sole purpose is to re-establish control of its designated OV should the vessel itself be overcome or commandeered by enemy forces.

For this reason, all MoD employees are urged to take great care in compartmentalizing both information relating to the physical location of the failsafe, and the hashkeys used in maintaining the communications link with its OV.

It is vital to the operational security of MoD, and that of the worlds our infrastructure administers, that no sensitive data be committed to memory, written down, repeated to a family member, or discussed with anyone other than members of staff with a security clearance higher than your own.

Kind regards,

//Ombudsman/MoD/BoD
Earth Aerospace
2151.09.17.08:32

small time

.Fortaleza Residential.GGW/EH349—South America.2153.07.31.14:26

Otis parks his rented truck on the street by a stairwell fluorescent green and orange with local gang tags. The low-level employee housing inside GGW's fifteen grotto is separated from the slums outside corporate borders by very little, its porous walls having at times caused GGW to be deregulated and embargoed by Ombudsman and the alliances, although recently they have maintained good standing through stronger troop presence at the walls.

While Otis waits in the cab, engine idling, three men descend the stairs. He opens the driver's window.

"Frank?" asks the closest man.

Otis says, "Yeah. Which makes you Paulo. It's all in the back."

The other two go to the truck's rear, opening it up.

"Paulo's upstairs," says the man standing by his window, "he'd like to meet you."

Otis scratches his chin, watching the guys unload in his side mirror.

"I'm short on time, how long does he want?"

"Just a minute," says the man with a hint of Brazilian accent.

Otis tells it like it is

Otis and Paulo sit at a plastic table and chairs on the apartment balcony, smoking cigars, looking out at the brick shanty town sprawled across mountains on all sides of the valley they occupy.

Paulo asks, "Is Keo okay, health-wise?"

"Sure," says Otis.

"I ask because he sent someone in his place the last three times."

Otis puffs on his stogie, leaning back to exhale.

"Keo's just a busy guy."

"I don't like dealing with go-betweens." Paulo takes a sip of Walker, brow creased. "It's a dangerous time for us. Enforcement are cleaning up down here."

Otis says, "I understand."

Paulo turns to Otis and asks, "Do you? Or are you placating me because you have more pressing concerns today?"

A pause. "Honestly?"

The man who greeted Otis outside steps onto the balcony, leaning over Paulo to whisper.

"You guys fucked it," says Paulo. "You gave us a case of AA-nines instead of the tens."

"Keo explained it already. He worked something out with Morlando."

"Morlando's in lockup, Frank. He didn't mention a thing."

"Okay. I'll get Keo on call. You guys can discuss."

Paulo says, "This is exactly the kind of thing I'm talking about. A lack of communication that arises when people send underlings."

Otis looks from Paulo to the grunt, then back.

"You've made yourself pretty clear on that issue, Paulo. Nothing's fucked here, you're paying a lower price for the nines. We'll work out something else when we can lay a hand on another crate of tens."

Paulo stands at the table, pointing at Otis with the hand trailing smoke.

"Why the hell would we need an anti-aircraft rifle with no higher-wavelength tracking? What would we do with it, Frank? Or is it Otis Wilmington? I've heard conflicting reports."

Otis sniffs. "You need to chill, Paulo. This isn't going anywhere good for you."

The grunt's hand drifts to his sidearm.

"Explain to me," says Paulo, "why I should eat the cost of weapons we can't use or sell. Give me a reason to let you walk away from here with our funds."

"I'll give you three." Otis tracks the grunt peripherally, annoyed now, refusing to break Paulo's stare. "One, you pull this shit on me and you lose access to the sixty-one arms dealers we ally with overnight. I think that covers most of your suppliers, mainly the specialty stuff. Some cardiovascular mines would be helpful when shock troops come rolling through your part of town, I'm sure."

Although he doesn't physically back down right away, Paulo is clearly aware of what has just occurred within the balcony's narrow social hierarchy.

"Two, a timed q-mail containing the specifics of our business relationship and drop locations—including this one—will be forwarded to GGW Enforcement unless I send a cancel key within thirty minutes. Standard operating procedure, comes in handy when some fuckrod grows an overinflated sense of self-worth and tries to lowball me on a trade under threat of violence."

Paulo slowly takes his seat again while Otis stares him down, cold as ice.

"And the third?" Paulo asks quietly.

"Three," Otis tells him, "is the fact that you bring henchmen with pistols to an arms deal . . . "

Otis opens his right hand, showing Paulo the sweaty, plastic detonator.

" . . . while I come strapped with enough composition to bring your whole world tumbling down around us."

Paulo and the grunt watch Otis for a long time, neither man moving.

"Never be afraid to lose more than you're willing to take."

Standing, Otis presses his cigar to the ashtray.

"Gents," he says with a sharp nod, before leaving.

measured response

. Mar del Plata Residential. GGL/SH59—South America.2153.20.21.14:26

Imogen curls up beside Otis on the couch, his arm around her. They watch the new Cleveland Spire film with the lights off, on the 79th floor of his apartment in Google's Fifty-Nine Shareholder sprawl.

An alert appears beside the main display, 'Buzzer!'

"Pause," says Otis. "Answer it."

The window expands, showing live footage of the building's front gate.

On display, six men mill about the entrance, their sleeve ink the same as graphics Otis spotted in the Fortaleza stairwell. Another man backs up a sports truck, the rest chaining the gate to its rear fender.

Otis says, "Call Enforcement," then turns to Imogen, earnest. "If they get through the gate this building is protected by blast doors and an eight man security detail. If they get past that, I keep a small armory here in the apartment. You're in no danger."

"No problem then. Who are the guys?"

"Disgruntled clients. I pissed off one of their execs yesterday. Now it's gonna be a thing I guess."

"Are all of your clients this violent?"

"Yeah, but most have better impulse control."

On display, they watch the driver floor it, the gate leaning, then bending open.

"It's a shareholder district," says Otis, "Enforcement come quick."

The men all react as gunfire starts coming at them from the building's lobby. They scatter, two of them taking cover behind the truck, firing back.

Security's sniper takes them both, a pair of shots that bore holes clean through the truck.

As the guards move in to confirm their kills, Enforcement cruisers set down around the dead men's vehicle. The driver, still inside, raises his hands when officers surround the car, rifles aimed.

"Well . . . good for us." Imogen takes a quick sip of his cocktail.

Otis says, "Resume."

Cleveland Spire slides down a rain-slick glass ceiling, shooting back at the masked gunmen who pursue him on flying skis.

the routine

"I'm sorry to wake you," Keo says on call.

Otis holds the PDA away from his ear, squinting to see the time. Beside him, Imogen stirs without waking.

"Hang on." He rolls off the bed carefully, going into the living room, closing the door. "This better be pretty fucking important dude."

"It's . . . strange. Earlier tonight an asset of mine in Cordoba called to set up an account on behalf of a new client. He told me the guy has deep pockets, says he wants to pay a tenth over asking price on a crate of CV mines."

"Okay . . . " Otis stares at an unopened pack of cigarettes on the coffee table, eyelids heavy.

"So the thing is . . . the client specifically requested that I involve you."

Otis sighs. "You're kidding."

"Also noteworthy is the fact that this particular asset was the one who originally put me in touch with Paulo."

"What the fuck is Paulo's *problem*, man? Could he honestly be this ham-handed?"

Otis grabs the cigarettes, going out to the balcony.

"This is what I'm thinking. It seems too straightforward. We're supposed to believe it's gonna be his play."

"Makes sense though," says Otis, "the stuff from earlier didn't feel right. Anyone with a basic knowledge of Enforcement response times could tell it was doomed from the start. Those guys weren't the attack. They weren't even a warning."

Lighting the cigarette, Otis catches movement at the edge of his peripheral, losing it immediately in the spark.

Keo asks, "So where does the attack come from?"

When his vision returns Otis blinks at the grinning side gunner of a UH-60 Bluebird helo, chaingun ghetto welded to a rig on its cargo door.

Dropping the cigarette, backing away, Otis says, "Shit fucker."

check

Concrete piercing rounds plow through the apartment walls at chest height while Otis runs with his head low, zig-zagging to the master bedroom, where

Imogen stands beside the bed, eyes wide. He drags her to the floor while .80 caliber rounds chew through the outer wall and windows, wardrobes on the far side caving in, coughing shredded Roman cloth.

"Go!" he barks, scuttling back to the hallway. Imogen follows, also croc crawling.

A haze of fine dust hangs thick in the air as all goes still at once.

Outside, hovering feet from the torn-open walls, they hear the gunner start to reload.

Otis leaps to his feet, grabbing Imogen, fleeing the apartment.

In the main corridor people peer out from their cracked-open front doors, watching Otis and Imogen run by. As they spill into an emergency stairwell the distant pounding of automatic gunfire on concrete resumes.

The residents all close their doors.

mate

.Orange Transit.OR/UH034—North America.2153.08.04.17:19

On the Uberhighway bridging Orange and Toys R Us territories, unperturbed by rush hour traffic, Paulo sits behind the wheel of his beat up Mercedes, smoking hash from a pipe, tapping the wheel in time with garage Threestep.

Otis watches through the scope on an LR-40 rifle as Paulo picks up his vibrating PDA, mouthing, "Answer it."

Laying beside Otis on the thirtieth floor rooftop, watching Paulo with binoculars, Keo says, "Hey Paulo."

On speaker, Paulo says, "Ah. I've been trying to reach you."

"So it seems."

The windows flash crimson.

While Otis dissembles the rifle, packing it piece at a time into his shoulder bag, Keo returns to the twin engine pleasure craft, starting it up.

Otis won't tell him that it's his first kill. He no longer has a sense that he can trust Keo, even a little bit. He sees the moment as an opportunity to strike out on his own in a sense, to become emotionally independent of his mentor, with confidence in his own reach.

The ability to underwrite word with action. It's a small victory, but he has drilled himself to dwell on these.

Musing on it, Otis takes the last of their equipment at a jog back to his Bosworth, eyes meeting Keo's briefly through the cockpit window, sensing now more certainly, the nebulous sinister leanings in his partner.

an unanswered question posed by time

.Dakota Residential.BP/AH4563—North America.2153.09.09.15:12

Imogen watches Zorena, doubt creeping in.

Zorena, with closed eyes. With the deep sea squid music and hanging windchimes. The alleged psychic abilities.

What a cliché.

"You've been having strange dreams," says Zorena.

Is there anyone alive who doesn't say their dreams are strange? "Right."

"You'll keep a dream journal," says Zorena, "mention everything you recall, the moment you wake. Dreams don't last longer than a minute in the waking mind. Over time, you'll see patterns."

"There's nothing more specific? Any advice, or things to be aware of?"

"Most people dream of their own future. The memory fades so quickly that they write it off as déjà vu when the event comes around."

Imogen sighs. "Yeah okay. I'll keep a journal."

"With the journal you see proof. It's all recorded."

"Yep."

Zorena opens her eyes, studying Imogen's face a moment. "Recently I dreamed of a woman who I believe to be you. I watched her be destroyed not through any machinations of her own, but the cloud of intrigue that fed off her circumstance, ultimately consuming her too."

"Wow," says Imogen, growing frustrated. "I don't even know what that means?"

"I dreamed of planets colliding. A man and woman who tore each other apart. An almost sun that filled the sky."

"Okay. Can you interpret it for me maybe? I have a hard time understanding other people's dreams, probably even more than my own."

"We all peer over the horizon in our distinct way. Write down your dreams. It's your bridge."

Imogen stands, smiling politely. "Alright, thanks for your time."

"I don't do lottery numbers," says Zorena.

"I didn't . . . "

"That's not how it works."

datalog

.Montana Operations.LM/OPS01—North America.2154.02.22.09:10

"This isn't official yet. There are no allegations of impropriety being made. We're just . . . interested to know why you'd seek out those directories in particular."

Imogen looks over the index on her PDA, confused.

"I honestly don't remember even looking at these. I can't think why I would. I'm sure you know about my condition, but it doesn't usually result in memory loss."

Benj Phuller tilts his head, intrigued. "Chronic fatigue, makes you fall asleep all the time?"

"Yeah."

"Do you sleepwalk? Wake up in new places?"

"No. Everything just blurs together. There's no clear separation between days or events. If anything I remember better, to me it all happened yesterday."

"So you recognize nothing in there?"

She shrugs. "I'm sorry."

"It's okay," says Benj, "I may need to follow up with you at a later date but for now we're good."

are you trying to break my heart

.Montana Residential.LM/SH05—North America.2154.02.22.19:04

Otis enters the apartment with flowers, walking straight to Imogen. He catches her off guard with a long kiss, leaning back to smile.

"Okay this is weird," she says, gently pushing him away, "what's happening?"

Otis turns, shrugging off his coat. "I was doing some tax stuff and noticed that it's been a year."

"Since?"

"When we met."

Confusion turns to slight amusement. "Wow."

"So you wanna go get some dinner? Somewhere nice."

She hesitates. "I guess."

"Are you tired?"

"I'm okay today." Imogen walks to the bedroom, looking through her closet. "That's funny," she says, standing behind a dark blue dress, absently watching herself in the mirror, "a guy at work was asking me about that night. The date didn't even register."

From the other room, "Yeah?"

"They think I was looking at some flagged files. I don't remember where I . . . "

Imogen returns to the living room, Otis fixing himself a drink in the adjoining kitchen.

"The day we met . . . "

He looks up, showing her the bottle. "Want one?"

"That was the same day Lewing Enright died."

Otis raises his glass. "To Lewing."

"I didn't log in at all that day."

"Log in to what?"

"The Lauchlap Michael PN."

Otis looks at her for a long time, expression fixed, before taking a sip of Tom Collins.

"You came over that night. You rented a car and drove me home, then you stayed."

He sips again. "Yeah. What a magical evening. Ready?" Otis finishes the drink, picking up his PDA.

Her eyes give it away, the rapid succession of logic flashing on her features as she puts it together. The scenario he anticipates every day, and somehow it still snuck up on him.

"Did you . . . break into Lauchlap Michael's network using my login?"

"Pshhh," says Otis, making a face. "Why would I do that?"

"I don't know. Why?"

Laying the impatience on thick, "Are you ready to go babe?"

"I'm in my pajamas."

"Oh." Otis pours himself a glass of strong gin, avoiding Imogen's bewildered stare.

"Otis did we meet on that street corner by chance? Or . . . did you arrange it?"

He thinks about it, drinking the gin, eventually looking her in the eye.

"Would it be any less significant if I did?"

Imogen lets out a chuckle, deranged sounding, even to her. "Why?"

"I needed access to Lauchlap Michael directories for a client."

"Then I was a job to you?"

"At first. Obviously not after the sex, and dating you for a year. At some point it just kind of fell away, all the beginning stuff."

"Otis . . . "

"I didn't know you then."

She walks away shaking her head, locking the bedroom door.

Otis looks at the empty glass, wondering if he should leave.

dreams are always wrong

.Montana Residential.DDG/CM03—North America.2155.06.20.20:21

Imogen walks alone by the harbor docklands, noticing the smell that hangs thick in the air before rain. She wonders why there are so many couples and families out at this time of night.

The fireworks don't startle her, so much as bring her back to reality. The crowded boardwalk glows red, then green, red again. Everyone looks up at the sparkling sky, transfixed. For a moment she wonders if another new years eve has crept up under the radar.

Over calm water, pinwheels begin to spin, shooting fire.

They form the shape of a massive giraffe, towering above the yachts and pleasure craft that have turned out in force. As the pinwheels die out a hundred meter high display rises from the water, revolving.

It reads, '2154.06.20'. Someone blows an air horn.

Simulcast music blares from PDAs amongst the crowd, ramping up, its beat timed to the fireworks.

A hand painted banner hung across the awning of Dodge's embassy says, 'Amsterdam Day 2155.'

She hadn't even considered that this would be a thing. Imogen is struck by the awareness that she's been alone now longer than she was with him.

warning sign

.Isidis Accommodation.WF/LH03—Mars.2159.10.02.02:06

Apple knows that she's drunk, that she shouldn't go up there, but it's not enough to stop her.

Let tomorrow worry about that, she tells herself, slumped against one side of the elevator. Fuck it all.

On 780 she stumbles out, struggling to read numbers on the doors. At Sylber's room she knocks. Then tries the bell. Knocks again.

He answers half asleep.

She kisses him and he reels back, more awake now.

"What are you doing?"

"There's been this tension between us since we met," she tells him, trying not to slur.

"Apple . . . "

"What? You don't like me?"

Sylber rubs his neck.

"I'm gay."

Swaying, she frowns at him, unsure if she heard right.

"Are you saying that because—"

"I'm saying it because I'm gay."

Apple takes a lumbering step backward, out into the hall, looking damaged.

"I'm sorry, I thought you knew. I didn't mean to . . . lead you on."

She turns away, power-walking back to the elevators with her head down.

Watching her disappear, Sylber makes a mental note that this is the time their balance shifted.

debrief

Hallen Wilde sits at the round table in a borrowed hotel conference room, eating cereal while swiping through a local tabloid.

He looks up when the door clicks, watching Apple and Sylber enter. They sit opposite, both placing their PDAs on the table.

"Interview with Hallen Wilde," says Sylber, "regarding the events of September twenty-nine, twenty-one fifty-nine."

Hallen looks at the PDAs. "You're recording this?"

Sylber says, "It'll be classified under company statutes. No one outside of us has access to it."

"Can I be prosecuted?"

"Not on Earth," says Apple. "That's where we're going to send you."

Hallen seems unconvinced.

"We've got your back," says Sylber.

Keotis

.Bolivia Transit.OM/4055—South America.2155.08.07.10:41

Otis runs through the crowded airport to escalators that rise into the upper sky lounge, bouncing off the occasional traveler as he ascends them two steps at a time. At the lounge he scans the room, spotting Keo watching basketball in the far corner.

Otis flops down opposite Keo and says, “Okay you want to hear this now?”

Keo continues to watch the game. “Shoot.”

Otis waves over the waitress, ordering a Jameson.

“When I was in Fortaleza,” says Otis, “Paulo referred to me as your employee. He wasn’t far off. If I let you in on this, if I return the courtesy you showed me in sharing the MoD stream . . . then we’re partners. Right down the middle. We form an alliance and go fifty-fifty, from today.”

Keo looks at Otis now. “What if I’m the one who lands it? I’d be giving up a lot of cash.”

“And what if I find it?”

Keo ruminates. “Yeah . . . okay. Lock it in.”

Otis throws back the scotch, quietly pleased with himself. He waves at her again.

When she’s gone Keo asks, “What’s the lead?”

“A woman on Titan who says she knows where it’s being built.”

Keo laughs, loud enough that people look over. “They’re not building it.”

“Why?”

“Where would they use it? I’m pretty sure there are no cities on Jupiter.”

"Maybe it's a planet outside the system. Exploration is booming right now . . . what if they found something worth settling?"

dragon's tail

.Xanadu-Regio Residential.OM/THC788—Titan.2155.08.19.08:34

☢

Parked in their rented sedan, Keo and Otis watch a four bedroom house amongst suburban spread west of Hobart city.

"Too open for an ambush," says Otis.

"Maybe an ops team in the ceiling space or garage. Can we get a look in that side window first?"

Otis opens his door. "If Ombudsman or MoD had us on their radar we would have been intercepted at Quarantine." He gets out, crossing the street.

Keo follows at a distance, eyes scanning the orange clouds for movement.

admin

☢

In the kitchen, Monique Zenith prepares bagged lunches while Otis and Keo watch. Her husband and son rattle about upstairs, a seamless morning routine.

"It's not something I was ever supposed to talk about . . . there were lots of memos. Binding agreements. They were pretty specific about keeping things under my hat."

Keo snorts. "Now that we came all this way, you want to renegotiate."

Monique looks up from the pita she stuffs with diced lamb, ardent. "What I know is worth it to you. To anyone who gives a shit about the vessel for whatever reason. MoD have gone to great lengths to seal it off, even meeting you puts my family at risk."

Keo sighs.

"My husband works. He's paid well. This is something I can walk away from and feel good about it."

Otis says, "If your info checks out we can go as high as three mil."

She wraps her son's lunch, labeling it. "I place its value closer to five."

He doesn't hesitate. "Five then."

"The OV you're looking for," says Monique, "was codenamed 'Mobius' by MoD."

Keo shoots Otis a look, giving nothing away.

"The project was kicking around for years, under the radar. MoD shopped it to every major aerospace outfit but no one would touch it. Even Lauchlap turned it down twice. Their Relations people said they saw no need for a platform the size of Mobius. Then suddenly, pretty much overnight, that changed. Mobius was fast-tracked as a joint venture, renamed Behemoth and shuffled onto Lauchlap Michael's shadow books. Construction began in secure space four years ago."

She finishes her coffee in one big gulp, rinsing out the mug.

Otis asks, "And somehow, you know where."

"It came across my desk one night. I was an exec based in Luna city, not cleared to see it. We reported the error and MoD cleaned house, got all of our d-sigs on an NDA and disbanded the department. They sent me here, and let me run an admin district in town."

"Then you're . . . what?" probes Otis. "Disgruntled?"

"No. I like my job. I like my life. My extended family are all here too."

"Why share this?"

"Because you're paying. I don't need it now, but it can't hurt to have a rainy day fund. I either use this leverage while I have it or lose it when they move Behemoth to another anchorage for supply and armament."

Keo says, "Okay. Let's hear it."

"Let's see five million Euro."

always be closing

.Jackson Wiltshire Residential.PRL/FC09—Ganymede.2155.08.24.16:20

Departing the large fishing town of Archimedes, Otis and the Mungtao reps meet at the ferry's bow, wind whipping at their winter coats and scarves.

Pleasantries out of the way, Otis addresses the eldest with what he hopes is a demeanor of respectful resilience.

"In the time since we last met, I've been made aware of the intrinsic value of Behemoth to a firm such as yours. I'm also up to speed on your ties to McPherson, loose as they may be."

The rep nods once, eyes fixed on his.

"I'm going to overlook the initial white lie," Otis continues. "And given my recent unrivaled run of progress on the matter, as well as my increasing costs, I'm going to renegotiate terms at this juncture."

After brief consideration the old man says, "This is acceptable."

outbound

.Hyperspace Biway KLTR-45635436775—Restricted Space.2155.08.29.00:03

Traveling at Light 6.1 aboard the Bosworth, Otis can feel Keo's eyes on him, aware of several sharp inhalations as if to begin speaking, before stopping short.

Eventually Keo spits out, "The best liars let you believe that you can tell when they're lying."

Fighting fatigue, an overall shitty mood, not wanting to engage in anything deeper than a polite nod, Otis looks over at him, eyebrows raised.

"It's a risk . . . getting bogged down in loyalties with any third party."

"You don't think we should be in the business of finding it?"

"Not with Mungtao. Not at their prices." Keo folds his arms and turns away, loosening his straps, trying to get some sleep.

Otis returns his attention to nav displays projected across the canopy. He twists the holographic dimmer, fading the whole array, getting his first good look at deep space. The stars stand out brighter here against the fabric, and there are no biway markers. Not another ship in range. No light beyond the thin film of their hull except that which pours from the universe itself.

gunmetal giant

.Rawling Proving Zone—Restricted Space.2155.09.02.02:12

Fifteen kilometers inside the no fly zone, at minimum safe distance from detection by the fleet patrolling their grid, Otis and Keo are nervous. In the cargo hold of their cruiser, Keo uses a small feed interface to fine-tune imaging equipment hotswapped to their hull.

On display, a hundred thousand clicks away, part of an enormous aerospace production station snaps into focus, its wireframe assembly built around the core mass of what they can only assume to be Behemoth.

Despite anticipating what they would see here, both are awed by its scope.

"That's got to be it," says Keo. "Big."

"Zoom out some more."

Welding drones flicker blue across the orbital vessel's entire surface area.

"It's zoomed out all the way."

Otis goes to the port window, staring out at the brightest speck of light.

"You're recording, right?"

"Of course."

Otis slowly shakes his head. "Why are they building it now? What do they know that the rest of us don't?"

we manufacture both dreams and explosions

.ISIDIS SHAREHOLDER.SYM/SA05—MARS.2155.09.02.02:12

In the revolving restaurant atop Symtex's Five Hotel, a wide semicircle of tables sit conspicuously unused on one of the year's busiest nights.

At ground zero, propped up against the window, watching every district of Isidis glide by in panorama, Keo orders the steak while Goseph Mailer opts for a vegetarian entrée. Goseph tells them to put a rush on it.

"I'll need copies of everything," says Goseph. "Clones of all the imaging logs. And transcribe the conversations you couldn't get on audio. They don't need to be verbatim."

"What do you think this guy is gonna do? You have his contacts at Mungtao. Close in. Scare him off. If he catches even a sniff of Ombudsman on his ass, he'll sprint back to small arms and bombkits. Trust me, Otis spooks easily."

Goseph sips his coffee.

"We're going to hold off. We want to see where this is going. You're our eyes and ears for now. It's a low rent task, I understand. But it can be an inroad."

"I have your word?"

"No. Not officially."

Keo's steak arrives first, the waiter avoiding eye contact with Goseph while he sets it down on the table.

Cutting into it, Keo says, "To be honest, I don't see your angle here. You want to keep Behemoth under wraps. You know that Otis has already found it, that he's holding the location to ransom until McPherson agree to his price. You have the capacity to stop him now, before the horse bolts . . . and yet you're all sitting back waiting."

"Assets in your position are rarely privy to the entire board. It's better that way. It's less people for us to kill when things are settled."

Keo senses in Goseph Mailer the soul of a man who will devour any one or thing that should appear within his singular, laser focus ambition. Keo chooses to skirt the threat, live large on the beast's golden leavings. Always orbit, never land. Only an opportune moment of weakness would fell someone like Mailer. While Keo has perfected the art of being a guy who can wait around for his moment to brick the bigger man, he sees his optimal approach here as passivity. Either/or. Keo will kill the king, or get rich waiting. And of course if he doesn't do it, someone else eventually will. It's physics. So for now he shuts his mouth, toes the line with Mailer and lets him take his entrée to go without a hint of smartassery.

As Goseph stands to leave, he gives Keo a look, a locking of eyes that says they both understand the kind of dynamic they will share in days to come.

hold on

It's been seventeen days, but it still dominates most of her waking thought. Apple can't help feeling that it all rests at the feet of Sylber.

Now that Pryce has corporate sponsorship, the game has clearly changed. There's no longer any chance of it being a whitewash.

On her display are Ombudsman procedural logs. Hundreds of thumbnailed directories, painting a bleak picture of their odds. In the history of Worldsnet Alliance, three thousand and thirteen corporations have run up against their affiliate members, economically or militarily. Four of those corporations still exist, in varying stages of bankruptcy.

It seems to be falling apart. This is the pivot point, where they either bring it home, or fail at what they had set out to do. Apple promised her million plus shareholders that Ennis Pryce would stand trial. She intimated that he would be executed if the primary goal was subverted by Ombudsman's systemic corruption.

She guesses that how she sees it now is the way Sylber must have felt, the impulse steadily rising to lash out in the face of overwhelming frustration.

reject

On call, Hallen hesitates.

In the back of her towncar, alone with the driver, Apple hangs on his next words.

After a long pause Hallen says, "I don't think I should resurface with things the way they are."

"You'd be safe here. Last time didn't go so well, I know. But with us you would stand a chance. We'll protect you."

"Last time I tried to kill Ennis Pryce," says Hallen, "he wasn't Worldsnet affiliated."

Sensing it slipping away, she rubs her chin. "What can I do to make this work for you?"

"Tell me the rumor isn't true."

"Rumor?"

"That he's focusing his newfound resources on putting me in the ground. He thinks I killed his dog or something?"

Her mind races for a fresh angle of attack.

"Apple, I'm sorry," says Hallen, ending the call.

irregularity

.Tampa Light Preserve.HIG/OB01—North America.2159.10.20.03:12

In the rotating upper dome of Higgins Observatory atop Mount Gale, overlooking Tampa Valley's volcanic nature preserve, Jack "Mozart" Davies reclines on the primary viewing deck's leather command chair. Looking up at stars splayed across a glass ceiling, he taps corresponding panels that magnify each in turn, offering comapartive wobble data and interpolative suggestions on extragalactic planet candidates.

An ethereal harmony echoes inside the big dome, audible frequencies of his nightly superspectrum survey of Earth's skies, translated to music on the fly by code that one of his students wrote.

"Can we go back?" he asks no one in particular. "Tenth of a degree."

The music stops. Downstairs, one of his team swings the dish back, the harmonics spooling up again.

"Yeah right there. Something poking out, behind Jupiter."

"We've been in mid solar shadow forty-two days," offers one of the techs.

"Right. What have we got on the other side?"

"Nothing. It's all MoD."

"We used to have about twelve relays around the moons, when were they retasked?"

Silence as his team of three search co-op records.

"A year and a half ago. All on the same day."

"Why?"

"It's off limits. They commandeered every neutral observatory beyond Jupiter."

"Let's try to get a visual on this thing. It's big. Can we access any relay logs from before the handover?"

"I'll try," says the same tech, "but if this is anything toxic it'll flag our search."

"That's okay," says Jack. "Maybe we'll get a call, and an explanation."

posse recruitment initiative phase B

.RIO TINTO TRANSIT.LFT/AP09—EU.2159.10.25.14:58

Eli Lylo searches the crowd for his driver, checking the airport's page service. While waiting there at the desk he sees Apple, pushing through other people's loved ones, waving to him. "Eli!"

"They told me they sent a driver."

"I was in the area. Want me to carry something?"

"I'm good," he says, following her to the parking structure with a single carry-on.

rogue trader

.Oporto Residential.AT/SH01—EU.2159.10.27.11:35

The vibrating PDA wakes Sylber with a start. It takes him a moment to get his bearings.

"Hello?"

On call, Apple asks, "What if we've been overcomplicating things?"

"What do you mean? What time is it?"

"It's late. Listen, I think I found a window. I can't go into it now, but if things go bad for me, I need you to know that I've offered Eli a sponsorship."

"That's fine, I like Eli . . . what do you mean if things go bad? Where are you right now?"

"I'm in the car."

"Going where?"

'Call ended by caller.'

"Fuck sake Apple," Sylber rolls out of bed, taking a dryclean-wrapped suit from the small closet, rubbing sleep from one eye.

stitch in time

.Utopia Accommodation.AND/SH88—Mars.2159.10.28.01:12

Exiting her cab at the front of Androgynon's twelve star shareholder hotel, Apple enters via the lobby, passing through security before approaching the concierge. She asks, "Do you know who I am?"

He nods. "Is there something I can help you with?"

"I need a roast turkey sent to my room. Whole. Quicker the better."

"Absolutely, Ms. Zenith."

He rushes away while she goes to the front desk.

"Zenith," she tells the girl there. "I reserved eighteen-forty-one."

In the elevator, her hands are trembling.

in the waiting

A pinprick camera outside watches the room across the hall, streaming live images to her main display. Apple sits on the emperor-sized bed atop its sheets, waiting. She eats a minibar Snickers, picking its nuts from her teeth.

The warped fisheye image of an empty hallway and quiet door, perfectly still for long enough to play tricks with her eyes, occasionally flickering when light changes outside its field of view, indicating movement from the elevator.

Each time she blinks away fatigue, scooting forward, waiting for his approach. Finally someone does drift into frame, hovers there at the door opposite. She can hear his clumsy, heavy footsteps right outside. This is the closest she has ever been to the man who killed her mother.

On display, Ennis Pryce sways, leaning on the door a moment before entering.

Beside her on the room service tray, Apple's turkey sits uneaten.

double back

Ten minutes are spent breaking the front door's encryption. Apple leans against the hallway wall attempting to look nondescript while her PDA churns through probability bonds. She finds caution unnecessary in the empty hall, the realization doing nothing to steady her fingers on the dimmed lightpad.

Tucked into her jeans, under the jacket, the carving knife is wrapped in a pillowcase.

When the display flashes green she cautiously lets herself in, creeping through the living room of his darkened executive suite, breathing quietly.

At the bedroom door she pauses, unrolling the cloth, getting a firm, comfortable grip on the knife. Ennis lays face down there on the big bed.

Coming around it, willing herself the mental fortitude to land two quick blows to the base of his skull, she spots the scattered painkillers, the empty bottles of cough syrup and sleeping aids.

Apple watches Ennis for a long time, unable to detect the rise and fall of his breath.

"Ennis?"

He doesn't react.

"Lights," she says, kneeling beside him, taking his pulse.

She lifts one of his eyelids.

"Call EMT," says Apple. "Possible overdose in this room."

what the sea wants

.Elysium Operations.WOR/OPS03—Mars.2159.10.01.09:09

Ennis is beginning to fit in well on Mars.

He attends the daily briefings with MoD liaisons, keeps abreast of what's going on within the system. Even watches the news.

People at Worldsnet don't know him as the Amsterdam guy. They don't try to know him at all. The friends and associates he meets, Ennis can walk away from at a moment's notice. And yet, his work and the life built around it are fulfilling in a way that nothing else has been.

He feels renewed. Different, in more ways than the tangible. Released of everything from before.

the planet Tyche

.Elysium Operations.WOR/OPS06—Mars.2159.10.04.17:04

After the briefing several key players are asked to remain, the room's privacy state activated when the last low tier members have filed out.

Ennis sits up, watching Kerrick Damon's expressionless face. Damon says, "Guys, as of this morning I'm authorized to inform you of a horizon situation that will more than likely impact your long term financials and logistics."

The assembled execs and sponsored ops exchange glances. Ennis can tell that this is unusual.

"The planet Tyche," says Damon, "is a rock giant native to this system, on a thirty-six hundred year orbit around our sun. Diameter a fifth of Jupiter's, and its mass exceeds any known object. We'll see that you're filled in on the scientific minutia in due course."

Long range imaging cycles through slides on the wall of displays behind him. Tyche is just another blob of light in the sky, identified only by accompanying graphics.

"Its location and vector will remain classified. What MoD *can* tell you at this point is that in less than a decade, Tyche's counter-clockwise path around Sol will intersect Earth's more conventional clockwise orbit." Damon pauses for effect. "We calculate a point-oh-four per cent chance that the two bodies will collide."

Some of the men are solemn. Some amazed. Most though, stone cold poker faces.

"Odds are six-point-eight-one per cent that Tyche's magnetic field, passing in close proximity, will trigger a polar shift on Earth, causing planet-wide devastation. That number is still somewhat sketchy so we're outsourcing the physics review."

Ennis is listening, but no longer here. Aware only that none of it will sink in until he's alone.

"And here's where things get kind of serious . . . we have it at just under twenty per cent that regardless of whether we experience a hit, or a near miss, Tyche's gravity will shunt Earth from its axis, sending it on a four year spiral into the sun."

For now, he just has to breathe. Let Damon finish up. Exchange customaries. Through the door, into an elevator. Maglev on seventy, the shareholder hotel three stops over.

Never let them see you sweat. Don't let them know you're scared. It will all be okay. His vision warping at the edges, every sound an echo, Ennis can think only of Amsterdam.

"We have nine and a half years," says Damon, somewhere distant. "And as yet, no one has put forward a cohesive idea on how to proceed."

physicists today announced conclusive proof that the flow of time is speeding up

Ennis feels that the world is moving by on fast forward, senses the divide growing between himself and those he encounters. The reality of continued detachment is now settling in his bones.

None of this will matter when Earth is gone. People will splinter. Countries will rise and disagree on every stellar patch of dust. War will go on because without it no one can ever be a hundred per cent right.

Any allegiance he has will be renegotiated when the power structure shuffles. All the people who stand to die in this scenario, and Ennis doesn't feel a thing for them. For anyone. Survival is all that matters. But survival isn't a feeling. He's left alone always with the constant, glaring obviousness of how those other parts in him no longer beat.

Twice a week Ennis flies back to Elysium, meeting with Planning & Infrastructure liaisons to MoD. Staring absently at their outcome reports, their guesstimations on future dead and dying. He can only wonder when everyone in every field became a so called liaison.

Earth has, they say, a 44.4565357377657% chance of death. The odds grow shorter with each face to face. After one briefing, with a day and a half to kill, Ennis drives out to his old ranch property, watching from a distance. Children play in the front yard. A swing set idles where he found Graeme dead, creaking in the breeze.

Pulling off the paved road, Ennis takes the "all" terrain rental onto a sloping field of red powder, aimed at distant mountain ranges, gaining speed.

Three clicks overhead, a Worldsnet satellite tracks his sojourn deep into Mars desert.

wanderlust

.Elysium Unpartitioned.OM/UP22007—Mars.2159.09.28.05:37

Ennis descends the canyon on its north wall, using the slippery grooves cut into spongy rock by running water. At the canyon floor where the water pools he stops, leaning over it to drink.

Standing again, Ennis looks around, his eye catching movement from the southern wall's shadow. Deer roam freely about the grassed fields, basking in morning sun.

Ennis walks east, into the canyon's depths.

dirt division

.Elysium Unpartitioned.OM/UP22009—Mars.2159.09.28.13:02

The Parks Department cruiser crosses the desert ten meters above it, setting down by one of three large rock formations. Ranger Buchannan exits the cruiser holding his hat against the rising wind, walking to the rocks.

In their shade he finds Ennis, unconscious, lips blistered and face red, a hodgepodge of sunburn and dust.

"Son?" He shakes Ennis, who stirs, squinting up at Buchannan.

"Yeah?"

"I'm the Ranger. Your people asked me to come and get you."

Ennis looks over at the cruiser, standing, brushing the dust from his clothes.

"My people? Okay."

He follows Buchannan back to the vehicle, defeated by the elements, thinking fondly of air conditioning and massage sofas, a renewed appreciation for the places he calls home.

intimacy

.Utopia Medical.WOR/MD14—Mars.2159.10.27.13:19

The Worldsnet psychologist tells Ennis that his visit is a formality, but he feels her sizing him up. Wondering if he's finally lost his shit altogether.

He says, "I'm like two different people. There are two personalities, both named Ennis, both aware of each other. I know how that sounds, but I'm being figurative."

"Define the differences between them," prods the psych.

"One guy is easygoing. Likes to laugh. The other one is logic based. Mature. Overly self-aware."

She asks, "Which one planned to blow up Amsterdam?"

Ennis pauses. "Are . . . you supposed to ask me about that?"

"Why wouldn't I?"

"Most people don't."

She doesn't respond to that, just writes.

"What did you write?"

The psychologist shows Ennis her display. 'No one asks him about Amsterdam.'

Leaning back, he asks, "Is that significant?"

She tilts her head, inquisitive. "I don't know. Is it?"

"For the record, I never planned to blow it up. I just wanted to go home."

She writes again. "If that were true, wouldn't you have disabled the bomb?"

Ennis watches birds fly by her window.

"It was plan B. There had to be a fallback."

"Kill everyone is your fallback?"

"I know. It was extreme. It was selfish. What more can I say?"

She puts her pen down. "It was more than any one of those things. It was abhorrent. It was unthinkably depraved. The truth is you never can atone for it, so the only way you'll ever get any peace in life, is to aimlessly hammer away at trying to make the universe a better place."

"And that's unhealthy, right? To be constantly moving? Pursuing an itch that I could never really scratch?"

"Is there a deed you could perform that would allow your conscience to rest?"

Ennis shakes his head.

"So it would be tempting to give up. To decide that you're hellbound already, why not just relax into the role of asshole."

"Sure."

"The only time you feel calm is when you're being proactive. You can channel that energy into further condemning yourself, or you can spend all your time steering as hard as you can toward redemption. I'm not an authority on religion or karma, but intention has to count for something somewhere, even if no one ever knows it but you."

Ennis stands, zipping his jacket. "Thanks. I'll come back next week."

"Ennis?"

He pauses at the door, turning back without meeting her placid stare. "Yeah?"

"What do you fear the most?"

"Hell," he says without thinking.

"What about here in the mortal world?"

This time he gives it consideration.

"Indifference." He opens the door.

hotel Androgynon

.Utopia Accommodation.AND/SH88—Mars.2159.10.28.02:53

When the RobotossDXM and Rite-2-Slepe is gone Ennis starts on a bottle of Mockphene, losing track of his intake over time. The ceiling warps in on him with each breath, light dripping from it. The bed envelops him, more than comfortable, but not forgiving. He doesn't sense his end. Doesn't seek it. He just doesn't care enough to exercise due caution.

What was started in Amsterdam, the universe finishes. There was never any other way. No loftier outcome was available to him, beyond that fork. Everything between there and here was guided by rails.

He sees now how fate conspires to end everything, and everyone, in time.

And how time is all it has.

bereft

Aboard the long haul passenger ship *Gallagher*, flying hyperspace biway Saturn-JK12, Umberto wakes in the twin bunk cabin he shares with Monique to find her gone. He pulls on a shirt, brushes his teeth, all the while debating whether to go looking.

The maze of hallways funnel him into the smoking deck and recreational grounds, a sprawl of restaurants and open "air" parks designed to make the weeks-long journey to Triton that much more bearable. The view takes him by surprise, mesmerizing stellar nurseries that linger in the distance.

He sees Monique talking to a man in his late forties by the wading pools. Maybe fifties. The man lights her cigarette, then his own.

Umberto spins on his heel, trots back the way he came, descending stairs that lead to the dining hall.

She catches sight of him there, an instant before he disappears.

For just that long, she tries to imagine a scenario where it ends well for either of them.

you miss my friends more than you ever miss me

.Newfound Finland Colonial.OM/CS82—Triton.2167.12.12.12:15

Umberto asks the counselor, "Would you risk your life for a friend?"

He thinks a moment, then nods. "I like to think I would."

"What if your friend started treating you like you weren't even there.

Avoiding you. Placating you. Making a conscious effort to cast you off."

"Then probably not."

"What if," pontificates Umberto, fingertips pressed tightly together, "the risk was of an ongoing nature, meaning the friend treats you like dirt while you continue to lay it on the line for them?"

The counselor smiles, reassuring. "Giancarlo, maybe it would be easier if you just explained the situation. What you say here is confidential, unless it involves a crime of some sort."

"It's not that clear cut," says Umberto. "Just humor me. If the friend started being for all intents and purposes a right *cunt*, would you turn her in? Turn *them* in. Thus saving your own ass?"

"I can't make this decision for you. You need to first and foremost discuss it with the friend. Tell her how you're feeling, but leave judgment and any strong emotions out of the conversation. Use neutral words . . . you remember we talked about those?"

Umberto notices the setting sun beyond Peter, the counselor assigned to him by Ombudsman's relocation benefits section on Triton.

the second girl

Victoria Transit.NV/RW7800—Triton.2168.01.01.02:01

Three floors beneath Victoria city on Triton, riding the light rail system between Flyweight Anchorage and Canberra, Umberto is going nowhere in particular on the endless city loop.

A trio of drunk teens share his rail car, loud but benign. At the other end, a young woman oblivious to all but the passing tunnel outside, and a businessman whose eyes occasionally dart to her, then back to the PDA he carries.

The car empties at Jackson terminal. Only Umberto and the woman stay.

From Jackson they ascend to street level, winding through the banlieues where sets of rails twelve wide are hemmed in by ad hoc arrangements of basic mud brick housing. In some of the windows, people watch the maglev pass.

Depression overwhelms Umberto at the thought of a person whose wanderlust extends only to the city circle.

"Are you from Earth?" asks the woman, still looking outside.

Approaching highways and taller buildings, the train rounds a tight bend, then begins the seventy floor climb to terminals that service the commercial crowd. Sprawled below them, the automated industrial district in 24/7 full swing.

"Yeah," says Umberto. "Do I have an accent?"

"Sure. But it's more about your gait, and the way you dress."

Umberto looks down at his shirt and pants, moves a knee to check which shoes he wore today.

"I've seen you here before," says the woman.

"I doubt it. I've only been on Triton a month."

"I see you riding around on the trains, late at night. Do you just do it for fun?"

"Sort of."

"Or are you avoiding something?"

Umberto flashes a glance at the 'Next Stop' display.

She says, "Maybe you don't have anywhere to go."

Pulling into the mid-tower Heineken terminal, Umberto stands, leaving the train.

On the platform he turns back to her, still looking away through the other window. He watches her glide off, the train's low hum fading as it rounds the IFC building.

Seven

monster

.Montana Shareholder.LM/SH02—North America.2156.01.07.20:35

Curiosity never used to get the better of Otis. Not with things like this. He could always compartmentalize the part that needed to know. It couldn't have taken control of him like this before.

Things are different now. Something has shifted in him. Hardened.

Already, he's severing ties. It comes too easily.

The first night she seems to sense the risk, always scanning her surroundings, taking the safe option of a walk by the shore with window shopping.

Imogen doesn't know he's there, watching from the rental, safe behind tinted windows and three lanes of traffic. But she smells something on the wind. Her intuition gives him away.

Or maybe that's part of the sickness. His paranoia-fueled fantasy world. The betrayal that exists only in his head.

Until he can prove it.

I shatter

.Montana Accomodation.HI/SA0034—North America.2156.01.11

The second night, she's less cautious.

Otis doesn't recognize the man. He's familiar to her however, and while there isn't any display of public affection between the two, their body language as they walk to his parked BMW is cause for concern.

Otis starts the rented Fordrollet, pulling into traffic at a safe distance.

When the car he follows stops at interchange onramp lights, Otis can just make out the passenger's silhouette, leaning across to kiss the driver.

this world and the next

.Elysium Residential.GP/EH982—Mars.2162.06.11.16:39

"Everyone, and everything, has a counterpart," says Weaver, "a metaphysical body that in some way represents them, in image and values. In a dream, if you spoke with me, you met my counterpart. When you dream that you're back in Amsterdam, you're obviously not there. The city streets you walk are a representation. Arranged the way you recall it. But does this place exist only in your mind, or is it an external thought form that you, even others, can access?"

Ennis lays with his eyes closed on the hemp carpet of Weaver's office, listening, focusing on breathing Qi into a ball at his navel.

Weaver himself sits with his back to Ennis, looking out across a digital representation of Antarctica, live feed from the continental reserve displayed across one wall—the small room's only light source. Weaver's face swells with awe, spills tears.

"On my wall is not the real thing, only a counterpart. I watch it. It watches me. On my desk, my wife's counterpart. My daughter's. It isn't them who watch, but through the counterpart they act. The dream them responds in every way the real them does. Not the actual them, but the closest them that I can have in this life."

Weaver seems to choke a little, sniffing hard, swallowing with difficulty.

"Any image or representation is a counterpart. But not all counterparts are simply images, or representations."

Ennis opens his eyes, sitting up, looking at Antarctica.

"Very rarely," Weaver says, "the counterpart is a vessel for the thing itself."

"What do you mean?"

"Sometimes my wife's soul speaks to me through her dream self, because the form is something she can comfortably manipulate. Sometimes . . . once in an eon . . . the thing on my wall *is* Antarctica. As if I could step right into it. Once in a lifetime Helena's counterpart is her, speaking through a dream, or apparition. The rest of the time, giving me guidance through the signs and synchronicities she inserts into our living world."

Ennis looks over at the unassuming desk, reassessing its minimalist appeal. He returns his focus to Weaver, who stares glassy-eyed, mouth agape, head tilted slightly back, at the place on his wall where smooth stone cliffs meet sea ice.

With an embarrassed cough, Ennis stands, putting on his coat. "Listen . . . "

Weaver looks away from the wall, but not at Ennis. "You're leaving."

"I thought you could maybe help me. And you did. But this isn't what I was looking for. Again, I'm sorry."

"Good luck to you sir." He returns to the image, which flickers as the door slides closed.

second chance draw

"Feed, Earth tourism. Something fun."

The shuttle jostles breaking Mars orbit, spilling his ginseng tea. Ennis orders another.

An unfamiliar lightness seems to grow on him, giving ordinary things a natural and humorous tone, a shiny appeal. He feels like a man who has taken the step after faking it.

There are well defined objectives ahead. A clearer code to live by. A stronger decision making framework. With his new found Zen calm, the constant state of flow permeating all moments, the one thing Ennis fears is losing it.

everyday actress

.London Transit.MG/HA044—EU.2162.06.25.13:01

Alexis Orrong sits in her idling Mini, touching up eyeshadow in its rearview while Ennis approaches on the passenger side, leaning in.

"Alexis?"

"Hi." She throws her tote bag on the back seat for him.

Ennis gets in beside her, filling the car with an aroma two parts leather, one part sweat. The word 'traveler' flashes on her mental canvas. It's the first time she's been aware of the strangely reassuring sensation it triggers, somewhere buried. The air of a man she knew? Maybe something more unique. The familiarity is frustratingly vague.

As they enter a CBD bound highway onramp she twists back to fish something from her bag, Ennis guiding the wheel when they drift. She comes back with a PDA, handing it to him.

"This is everything he had?" He presses his thumb to the dimpled scan on its translucent surface, generating a login prompt.

"Should be." Alexis guides them deftly between peak hour stragglers at twice their speed.

"Huh?"

"I can't login."

Ennis pockets it. "Where did you get it?"

She throws her head back, an exaggerated stage laugh.

"What kind of thief are you?"

She says, "The kind who doesn't talk about it after."

They drive a while without speaking, Ennis watching towers and sky cities grow taller, denser.

While night sets in and the streets begin to glow, lit from beneath, Ennis asks, “What do you know about Tyche?”

She shakes her head with a wry smile. “Tyche is bullshit.”

“Yeah, it sucks. For Earth at least. Do you think they’ll just evacuate the whole planet?”

“You’re not hearing me.” She turns to him, trying to assess the level of intelligence looking back. “Tyche . . . is not real. It’s a scam.”

Now Ennis laughs, only genuine. He taps the pocket with the PDA. “Is that what it says in here?”

“I told you. I don’t know what’s on there.”

The Mini takes an exit a click out from McPherson Tower One, descending into the subnet.

“What are you saying then, that it doesn’t exist? Why would someone fake a planet?”

She shrugs. “I wasn’t involved.”

slipstream

.London Entertainment.MG/ED12—EU.2162.06.25.19:40

“I heard you met with Weaver,” says Park, sipping Stella Artois.

Ennis nods. “Only once though.”

“He’s a good guy. Helped me through some stuff.”

“He wasn’t really what I was looking for in a therapist.”

Park smiles. “He’s a bit weird now. He lost his family a few years back.”

“I got that impression, yeah. I wasn’t going to ask about it. Any idea what happened?”

Park winces. “Oh. Right, yeah. His wife and daughter were in Amsterdam.”

Ennis feels suddenly sick. Not at the word, like before. Now it seems displaced. As if erasing that part of his memory has left a leak in the logical backtracing of where such crippling feelings originate.

Alexis returns from the bar with three more beers, leaning over the table to steady them by six empties. They all rattle as the PDA beside them vibrates. Ennis scoops it up, standing, feeling beer sweat coming on. “I’ll be back.”

He drifts between the other diners and drinkers, stepping onto a balcony before answering the call.

On audio, an unfamiliar voice asks, “Ennis?”

A mental rolodex of current friends and enemies flashes momentarily, long enough that the hesitation is evident to his caller. “Yeah?”

“My name’s Wes, I was with you in the Tyche meeting.”

Ennis looks around, at quiet families enjoying an outdoor brunch. He glances back at Alexis and Park, engaged in friendly conversation.

“Okay.”

“I’m calling to let you know that Kerrick Damon is dead. So are most of the Worldsnet capos who were in the room with us. Damon wasn’t supposed to share the information he did. Ombudsman never authorized MoD to disclose it.”

Ennis feels his chest tighten. “Why would they tell us?”

“The whole thing is looking like a soft coup by the Admirals. Then again, a colleague heard Damon was part of some rogue cell within Navy. Maybe a few of the hardasses inside MoD wanted to give people on Earth time to plan evacuations. It’s moot now anyway. DoDC are sealing it off.”

“Department of . . . ?”

“Damage Control.”

“So what? Watch my back? There’s a waiting list.”

A pause, Ennis can hear the familiar takeoff routine of a commercial shuttle service.

"Where are you Wes?"

"Worldsnet protected you until now, but there's a pretty good chance that one of their affiliates has been tasked with resigning you. I heard it was McPherson coming after me, I don't know who you got."

Ennis watches Alexis speak quietly to Park. A fork loaded with pumpkin lasagna hovers inches from her lips. It has stopped steaming.

Her strained eyes belie the wide smile.

"I think I'm starting to get the picture."

"Good luck then."

"Just tell me," says Ennis, looking up at the McPherson Group banner, floating above every building to the horizon, "is Tyche for real? Is it a big con?"

A low rumble on the other end. "I have no idea."

Ennis ends the call, watching Alexis finally take the bite of lasagna as he returns to the table with what he hopes is naïve amiability peppered with nonchalance.

He hates himself so much right now, for not having packed a gun.

ends

Ennis pauses before swiping into his room, turning back to Park. "Why would they fake a planet?"

Park looks up and down the hall, stepping closer. "It's a land grab."

"That's all?"

"All? That's a planet, man. It's *the* planet. They get everyone thinking Earth is going tits up, make it look like it didn't leak directly through MoD or Ombudsman's network. Hell, they even made it look like MoD leaked it without Ombudsman clearance. MoD will rebel, and be absorbed like every other body."

"This isn't just about an internal power struggle? It's too big."

"Of course not. When everyone evacuates, the big three come in and snatch up all the land, repartitioning Earth. Every sacred site. Every tourist trap. Every last piece of infrastructure abandoned by corporations in their panic not to be the last rats on the ship. By the time the smaller alliances and independents catch on, it's too late. Ombudsman is bringing in new tariffs, bribes, leasing fees. They have permanent strategic bases on every continent by the time everyone figures it out and returns. No action will be brought forward through the dispute system because they own it."

Ennis frowns, leaning on the doorframe. "That's a pretty fucking convoluted scam."

Park nods, turning back to the door across the hall, swiping in. "Of course it is. It's Ombudsman. How do you think they took power in the first place?"

Ennis stumbles into his own room, fading as he hits the bed.

escape from MG

In the morning, Alexis and Park are gone.

The concierge tells Ennis they left around ten, with instructions to go on alone. A company emergency, was all they could say.

Ennis takes a car service to the airport, playing chess against a guy on Titan most of the way home.

Hardly the retreat from McPherson territory he envisaged last night.

sociopathy for fun and profit

.London Health.MM/SHH01D—EH.2162.06.29.10:11

He hasn't even sat down when the Worldsnet psychologist begins reading from her display.

"Dear Amsterdam, as you may or may not be aware, I am a disgruntled former resident of your fine city banned from returning under penalty of death. My crime was setting foot beyond your borders."

Too familiar with the words, despite how alien they come across with her clinical delivery.

"Seeing as how I've got an A-Bomb now," she continues, "I decided to dictate this list of demands. I really feel—"

"Thanks," he interrupts, finding his voice, "I get it."

"This comes off as purposefully ignorant. It shows a sinister side that you *tried* to indulge, a diabolical little man you fostered with petty justification, stemming from what you perceived as extraordinary personal hardship."

He nods.

"You didn't *care* if one-point-two million people lived or died. It was a videogame. And levity made it that much easier for you. It turned the reality of their deaths into meaningless clouds of notion. You fell in love with the idea of being the guy who got back in or died trying, then found yourself with buyer's remorse when it all went wrong."

The scar tissue in his chest and back tingles. Ennis shifts in the chair, noticing his tightly clasped hands.

"You call yourself crazy bomber. Like it's a comic strip."

For the first time since Amsterdam, Ennis doesn't try to explain the sequence of events that led to his act of genocide.

She says, "So at the end of the day, it comes down to this. You either had yourself in such a state of denial that you were convinced these people all deserved to die, or, you're a stone cold sociopath."

He looks up at her, more ready to hear what comes next than he ever was or will be.

"Which one . . . " His voice cracks. "What do you think it is?"

"Honestly? I think you were always capable of what happened in Amsterdam. All it needed was the emotional trigger. You were able to distance yourself so easily because you don't process emotion or feel empathy the way ninety-eight per cent of human beings do."

Ennis shakes his head, scowling. "So that's just—who I am? I can't fix it?"

"I can give you medication, but it's not a cure. A borderline type personality is the hardest to diagnose, and to treat. The real question is, do you *want* to fix it?"

He snorts. "What?"

"Amsterdam was an unhealthy expression of what makes you special. But as a Worldsnet operative you've found a way to hone it, and of course in doing so, a healthy outlet for something that will always be a factor in your life."

"You're saying it's a good thing?"

Her eyes dart to the window, instinctively. Barely long enough, but he catches it. "Not at all. You killed a million people. But at some point we have to move beyond that. We need to understand that you are in fact a person with highly specialized skills. One of the two per cent produced by nature to keep the population in check. Worldsnet can provide a path to areas in which you would excel. You're good at what you do. We feel that you should keep it up."

"Wait, what? You feel that? Or Worldsnet does?"

She shrugs. "Both."

Ennis stands, absently walking to the door. "I have no idea what to do with that."

"I'll put through a prescription for you, pick it up downstairs. Things will look better with time and perspective. Please don't hesitate to call if there's anything we can do."

what to get for the man who has everything

At the Mars-bound departure lounge in McPherson Group's CBD one, Ennis is woken from a light Mirtazapine snooze by thunder, distant then close, fading in and out. He sits up when the screaming starts, taking a moment to register where it comes from.

Other passengers crowd around the lounge's tinted glass windows.

He joins them there, watching the last embers of a mushroom cloud turn to thick black smoke, pluming from the searing wreckage of two passenger haulers intertwined at the wireframe cores.

"Oh jeez . . . what happened?"

The woman next to him, already hysterical, shouts, "It just landed on top. Just landed right there like he didn't even . . . "

Ennis turns to one of the desk staff who has arrived at the window.

"How do we get to them?"

The clerk shakes his head. "I can't let you go out there."

Outside, fire and rescue trucks speed toward the debris field, their hoses trailing foam. A tanker helo lifts off in the distance, coming around, sluggish with its full reservoirs.

Ennis turns to the clerk, snatching his wrist, tearing the ID cufflink from his sleeve.

"Sir you'll only get in their way!"

But Ennis is already crossing tarmac, the exit door closing with a hiss.

body count

Through tears and smoke, he presses on blind.

His skin tingles. Hair stands on end. A man walks hunched over from inky smoke, holding his stomach, insides spilling out between fingers. He passes with a stunned nod.

Ennis continues into the next section, waving an arm ahead of him for stragglers. He trips on something soft and solid. In the darkness, the acrid smelling charcoal fog of catastrophic technical failure, Ennis can feel a face.

He takes the wrists, drags it back the way he came. A weak pulse beating against his palm. His chest burns more with each sharp intake of smoldering insulation. Ennis can see sunlight. A meter, maybe two, then clear air.

"Number one," he tells the unconscious passenger, on the verge of passing out. "First installment."

dogs are the best people

.Montevideo Entertainment.MT/KD28—South America.2156.01.12.23:27

El Chu Pa Ka Bra Asian Cuisine & Karaoke in Mungtao's semi-unallocated slums is winding down for the evening, with only the last few punters remaining at tables, watching the empty stage.

"I know we come close man, I know there are guys who are fan-fucking-*tastic* liars, but when it comes to partitioning your life to the point that your seventeen boyfriends and twelve-thousand best friends have never met each other, let alone the real you, those bitches are in a league of their own."

Keo watches a shark do circular laps of the inappropriately small tank that serves as the dive karaoke bar's centerpiece, its growth stunted, health poor.

For an instant he wonders how hard it would be to take care of a shark at home, snapping back to Otis when he senses the silence.

"Bitches are fucked man. Yeah." He sips his Bud, hoping the night will end soon.

But Otis is just getting started. He pours another shot of Sapphire, downing it quickly, then a second, and a third. He offers the fourth to Keo, who declines.

"Okay I mean, we're like . . . actors, juggling all the roles, having a cover story ready and all. But if you probe deep enough, we'll break down and tell the truth. It's not worth the shit storm I guess. It's ultimately cathartic. For women it's like, like a personality shift. They believe it. They've constructed a fucking universe around it that will hold up to the most intense scrutiny. Then they have *us* believing that we're paranoid, they become outraged that we would accuse them of something even though they know they did what we say they did. Like we're the fucking bad guy. And we say *sorry*. That's the worst part. Once they get that, they won't ever set you straight. There's too much invested. They'll take it to their grave."

He trails off, looking suddenly pale, leaning left, then forward.

"That's a surefire postmodern interpretation of the battle of the sexes. I couldn't have put it better myself. It's like you're the Dalai Lama of relationships or something bro."

Otis vomits on the table, sending Keo reeling back, disgusted.

"Okay I'm done." Keo flicks a chunk from his elbow, watching the barmaid get a mop.

Suddenly lethargic, Otis sways in his chair, moaning. "I feel sick dude."

Keo sighs, lifting him by the arm. "Let's get you a cab."

Otis nods, picking up the shot glass, throwing back two parts Gin, one bile.

As the woman approaches with gloves and mask, mop and bucket, he vomits again, this time on the floor at her feet.

Her nonplussed demeanor, that of a woman who regularly sees far worse.

prodigal

.Montevideo Accomodation.GH/SH10.2156.01.13.07:58

The day Otis quits drinking is the first day after deciding never to regret what happened with Imogen. To no longer dwell on it, or let it affect his movements in any way.

Breaking orbit over the North Atlantic, he sets the autopilot before going in back, collapsing on the Bosworth's cot.

The PDA in his pocket vibrates, catching him on the edge of dozing off. A tungsten ball inside his mind seems to roll against the sides with every jerk of the skull.

"Yeah?" It keeps vibrating. "Answer, yeah?"

"Otis?"

"Who's that?" His face buried in the pillow, Otis rapidly fades.

"Goseph Mailer."

Otis rolls over, taking out the PDA. Projected on its surface, the iDent reads 'Mailer, Goseph'.

"Shit," says Otis. "Um. What's this about sorry?"

"The orbital vessel Behemoth."

"Shit."

"Don't worry. This isn't what you think. But we need to talk."

"I doubt I can do that." His mind races for plausible scenarios to offer when Mailer asks the inevitable follow-up.

"I'm not asking."

A pause. Otis takes a shallow breath.

"Yeah. Okay, just . . . where? I'm kind of off-world at the moment."

"I know where you are," says Mailer. "And I know where you're going."

"Shit," says Otis again, seeing now how truly fucked he is.

"I assume you remember how to get here."

Behemoth's toe

.Location Unknown.MoD/CDD01—Deep space.2156.01.13.00:00

Mailer's aide shows Otis through lengthy corridors that wind at right angles through the ship-building facility at MoD's 'Clover' drydock.

Every ten meters, when they pass a window, Otis can't help himself. He always looks. The aide occasionally smiles at Otis, pleasantly phony.

As they enter areas marked by ascending security levels, armored sentries watch them pass, each with UVIR faceplate and a rifle Otis can't identify.

The aide leads him into a waiting lift, punching in the lower deck.

"Are we going onto the ship?"

"Not today," says the aide.

Ombudsman

"Our recent census revealed that people are growing uncomfortable with the delineation between Ombudsman and MoD."

Otis glances at the window again.

"As of Monday MoD and its navy will be reabsorbed into Ombudsman, a reinvigorated and newly focused organization dedicated to the growth of . . . " Goseph Mailer swipes his display. " . . . both the corporation *and* the individual."

"Wow," says Otis.

"Anyway, if you could put it out through your circles, we're trying to phase out the term 'MoD' by twenty-one sixty."

"Sure."

"Now let's talk Behemoth. We know about Mungtao's offer. We also know who sent them your way."

"McPherson?"

"Of course."

Otis shakes his head. "Fuckers. I knew it. I wasn't a hundred per cent, and those Mungtao douchebags would never say out loud, but it was clear."

"We're told that McPherson's aim is sabotage. Lauchlap have been on thin ice with us since the Longhorn fiasco, obviously they're looking to tip it beyond breaking point."

Still watching construction ships stream in and out of Behemoth, Otis asks, "So now what? If you know all this, why even bring me out here?"

Mailer punches tokens on his lightpad, bringing up a surveillance image of

Otis and Keo laying on the Mexican rooftop.

"Are you aware that your friend Keo has been reporting on your activities to MoD—*Ombudsman*—for several years?"

Otis watches the muzzle flash of the rifle he disposed of in Lunar desert.

"No. I was not aware." He mentally files the information for later, giving nothing away. Of course, in the effort to hide his microexpressions from Mailer, Otis has revealed more of his nature than he knows.

"I'll have our logs sent to your q-mail, I don't expect you to take anything at face value."

"Why would you tell me this?"

"Because you were never the target of our investigation." Mailer follows his glance to the window. "Would you like a tour?"

planetary

The bridge reminds Otis of various war rooms he's encountered, but novelty-sized. Alone with Mailer on the glass-floored viewing deck, he begins to wonder who would go looking if he didn't return from this trip.

He asks Mailer, "What kind of planet requires an OV this big?"

"Who knows? The point is to have one before you discover the planet. This isn't the only unallocated vessel being built in MoD space."

"Ombudsman," says Otis.

"If exploration turned up a habitable rock giant tomorrow, Behemoth would go into service the day after. All colonization would be done under our umbrella, circumventing the volatile 'wilderness' period other frontier worlds have lived through."

"If you don't find one? What's the shelf life on this thing?"

Mailer shrugs. "What's your point?"

"It seems like a colossal waste of funds for something that may or may not turn up."

"Finance and assets don't concern us at this level. Those come only as a byproduct of influence. Behemoth is influence."

Otis turns away from the window, looking around the bridge.

"So I know about Behemoth. I could lead others here. I assume you're either going to kill me, or offer me a job."

"We recently lost track of Keo. He killed some of our guys, slipped surveillance in Asia. We want to task you with establishing contact to report on his activities."

Otis laughs. "That's great. We can spy on each other. You can get our two perspectives and find the truth somewhere down the middle. It'll be like a Golf Renner film."

"Keo no longer works for us. I assume someone let him in on the fact that he was a person of interest to Ombudsman."

"Can I have some time to think about it?"

"Sure." Mailer stares at him a moment. "Well?"

wakeup call

"My dad told me to get in touch with you if anything happened."

"How old are you?" asks Ennis.

"Twenty. Why?"

"I thought your dad was a much younger man."

"Okay..?"

Ennis lifts his head to look at the time.

"Where are you right now?"

"Earth," says Mint.

"Me too." Shuffling on her end of the line, a loaded silence. Ennis doesn't quite know how to ask, so he just comes out with it. "What exactly happened to Wes?"

She hesitates. "His flight crashed on Mars."

second wave

.Beijing Transit.MT/MT4552—Asia.2163.08.02.14:26

Ennis Pryce and Mint Hamilton drink chai in the glass-walled underground hub of Beijing region's transport district.

"So you're . . . Amsterdam guy?"

"Yeah."

She scans the busy café, never looking him directly in the eye. "My dad had information that gained him some sort of leverage with MoD.

He left a copy with me, and his lawyer. The dead one."

"He told you to give it to me?"

"If things went bad."

"Why?"

She shrugs.

They swipe PDAs, the transfer confirmed by his.

"You know you're the second person to randomly contact me with secret information this month?"

"Lucky guy." Mint goes back to the counter while Ennis begins to read.

escalation

.Beijing Accomodation.FS/SH1—Asia.2163.08.08.04:21

At around twenty past four Ennis wakes on the couch with early news playing on the feed.

It's cold. He sits up, unsteady, willing himself the mental fortitude to stand.

The anchor's overly-concerned tone catches his ear.

On display, a mushroom cloud.

Another fucking Amsterdam day. Already? Jeez.

"Feed off." He stands, determined to avoid the day's media cycle.

Amsterdam II

And so, twelve hours later, Ennis watches news of Islamiyah with the same sense of astonishment that he assumes people felt right after Amsterdam.

In the unusually empty maglev car, Ennis is seeing the highlights of amateur footage that has been flooding into studios since around three this morning.

On display, the impossible spectacle of multiple mushroom clouds towering over the Great Dividing Wall. Satellite images of what will one day be known as the first continental carpet bombing. Security footage from inside the wall, chronicling the flashes, the fires, the dying.

He moves to exit doors, keeping his back to the display.

the fifth night

.Victoria Transit.NV/RW7835—Triton.2169.01.04.01:07

The third night after meeting, Jess and Umberto discuss her inability to get an off-world visa.

The fourth night, it's politics. Before they part, a brief rundown of his childhood, what brought him to Triton, only devoid of any mention of Monique or the circumstances surrounding her, which leaves a lot of space to improvise.

Tonight there are more pauses in the flow of conversation. He wouldn't call the silences comfortable. He wouldn't call them anything. Upon trying to analyze what hangs in the air between them, Umberto is confronted with the notion that it defies analysis.

On their twelfth tour of Victoria, sun rising outside, Umberto can see his station approaching.

Hesitation as he stands.

"I figured out why you do this."

She looks him up and down. "Yeah?"

"Yeah."

Jess watches him leave. She wants him to look back before descending the stairs.

empty train

The next night Umberto rides the mag-lev until dawn, scanning every platform for her. Double-checking faces. Jess never shows. Exiting the car, he tries to ignore the depressing idea of another day spent sleeping.

Umberto wonders if there's a better way to avoid running into Monique at home.

authentic

.Victoria Residential.BH/EH2600—Triton.2168.05.14.08:52

When he hears her leave their two bedroom apartment Umberto finally rolls out of bed, headed for the kitchen.

Preparing toast, he strikes a sudden realization, running back to his room. When he returns, Umberto tosses his PDA on the countertop.

"Call new user, wildcard, Jess or Jessica. Victoria city on Triton."

The PDA displays, '1,110 entries found. Call first user?'

"Call," he confirms through a mouthful.

'Calling Jess Aardvark.'

"Hello?"

"Jess? It's Umberto, from the train."

"What?"

"Never mind."

'Call ended by you. Call Aarons-Finch, Jess?'

"Call it."

unsent letter

.Havana Entertainment.KJ/CD0101—North America.2169.02.09.06:19

Exiting the casino via its rooftop door, Otis is surprised by the morning sun, having assumed even on his way upstairs that he'd sat at the table only for a few hands. He pulls a faded aviator's jacket tighter around his chest, folding his arms tight across it while taking long strides to the row of docking mounts.

When he boards the Bosworth there's a missed call from Imogen flashing in the cockpit. He lingers on the icon, not sure what to think. The last time she would even take a call from him was years prior.

A cold sweat. He lifts off from the charred concrete pad, setting course. Steering out between waiting haulers circling above stormclouds, Otis decides the message can hold until he's back on Earth.

But then of course, he plays it. In the empty spaceship, her familiar voice echoes. For the first time in as long as he can recall, Otis aches at the distance now between them. If it's a good or bad thing, he wouldn't know, but it makes him wonder if there might be a chance, sparking something kind of exciting behind rusted-on layers of acceptance. Not like the old days, but maybe to be something different. Same actors, new parts. Something along those lines.

fortune favors

.Montana Shareholder.LM/ST14—North America.2169.02.15.17:00

Otis gets halfway out of the car, the ringing PDA stopping him. He settles back into the driver's seat.

"Yeah?"

"Otis?"

More urgent, "Yeah?"

"It's Ennis."

"Ennis who?"

A pause. "It's Ennis. You know me."

Otis looks at the house, then the parked cars on her street. "How did you get my name?"

"Listen, I need to ask you for another nuclear weapon, bigger the better."

Otis laughs. "This is a prank call, right?"

"You remember what happened after Amsterdam. These people whose lives I fucked have all been baying for my blood nearly fifteen years."

"Okay..?"

"The greens and libertarians would love a shot, I'm sure, at the arms dealer behind the arms. The supplier as it were."

"Oh fuck off. You're out of your element, dickhead. Fucking . . . amateur. What kind of shit is this gonna be?"

"The men in power who stuck their neck out for me, that first time around. All these years later, do you think they would sacrifice more press time and political will to Amsterdam? Or this time around, for someone of your standing in the community, are they more likely to yield for Apple fucking Zenith?"

"I heard that she left that outfit."

"And I heard that her leaving was just more bullshit to draw me out."

Otis asks, "Why would I help again? You killed all those people for no reason."

"I had reasons."

"For an insane reason."

"Because this time I'm going to use my trademark weapon against a genuine bad guy."

"Remember how you're a mental though? Maybe you think the Unipope is bad now."

"Look, basically, I can circulate a rumor through AP and Indienet that there was a second mastermind, another guy who might have actually pushed the red button."

"People aren't stupid. No one will buy something that contrived, this long after the fact. You're a public liar, a mass murderer for God's sake."

"No—famous liar. Famous mass murderer. Personally, and this is just me kind of blue skying on the evening talk circuit, I find it a little odd that a lone madman shot four times in the spine was able to detonate a nuclear warhead as he fell unconscious to the ground."

"Why are you doing this?"

"Because no one else is."

energize

.Hawker Manufacturing.DU/MFD01—Titan.2164.07.07.20:26

In the slums outside Duracell's manufacturing quadrant on Titan, Ennis has been concealed in his camo tent eleven, going on twelve, days. This, at last, is his time. There is no apprehension left in his tired bones.

The flares drop on time, a little short of the target, but it will all be moot soon enough.

The orbital assault begins next, an array of three Ombudsman rails bombarding the site with enough kinetic mass that a half mile-deep trench remains where reinforced factories stood moments ago.

While the fires really begin to take hold, Ennis stands, arching his knotted back, cracking the neck.

In the valley below, Duracell burns.

"Splash one MFD," says Ennis into the open mic.

His ears pop as the approaching helo's stealth kills the distant wail of sirens.

pacifier

.Hawker Accomodation.DU/SH45—Titan.2165.07.08.21:04

General Gannon isn't listening. "It sends the wrong message," insists Gannon, on call, "my sergeants pitching tents at Gallagher Stadium while enlisteds live it up on the thousandth floor of the Grand."

Walking to the end of the hall, Ennis stops at a window, checking the weather. "I see your point."

"I've arranged camping grounds for your men just shy of the line. I need them ready to move when we are."

"You're not hearing me, General."

On audio, Gannon sighs, then ends the call.

Standing in the doorway of 1,103, an infantryman nods to Ennis when he passes.

stand down

At the end of the day, hostilities wind down without open conflict.

Relations remain terse between Ennis and Gannon's office.

Shipping out of Titan, bound for Mars, the Ombudsman Kelvin class carrier is hauled through anchorage yards by a pair of TUG engines, swinging the big vessel's nose at Sagittarius.

In the boost chair behind a four man flight crew, Ennis seems relaxed, the calm broken by his inquisitive-looking comms guy Tywon.

"Sir?" He hands Ennis a PDA, displaying a crawling transcript of communications between the carrier's command room, and Ombudsman's newly-established Mars headquarters. "Twenty-one oh-four, eight seconds in."

Ennis scrolls back. "Are you authorized to view these?"

Tywon shrugs. "No one told me not to."

On display, 'Marduk 1 reports internal heat source.'

Ennis looks up at Tywon, blank. "Okay?"

"Now go to twenty-two twenty . . . fifty seconds."

'Party reports surface inconsistencies.'

Ennis offers him the PDA back.

"Now twenty-two twenty-nine, thirteen seconds."

On display, 'Communication lost with Marduk 1. Communication lost with landing party.'

"What does it mean?"

"Three years ago you gave me a list of words to grep. *Nibiru, Behemoth, Mobius, Tyche*. Ring a bell?"

"Yeah? So what's Marduk one supposed to be?"

"Well Marduk is Tyche."

"How's that?"

Tywon leans, getting more comfortable. "It was known by a lot of names in the ancient world, but in each instance it's a supposed twelfth Solar planet on a thirty-six hundred year orbit around the sun."

The crew complete their checklist, lighting the engines with a thud that reverberates through the bulkheads.

"Well," says Ennis at last.

"Hm?"

"All of that was a long time ago."

The OSC Philadelphia moves away under a hard burn, its EM core producing aurora effects across the bow as it transitions to hyperspace.

"Probably just the name of some random ship," says Tywon. "But I thought you should see it."

slow

.Earthbound Biway.OM/EB2678—Shipping Space.2165.01.16.00:21

Ennis is vaguely aware of the morphine wearing off. The mess hall is empty, outsourced cleaners just now removing all trace of the twelve-thousand men who came through here on the way to the latest in a long series of situational briefings.

Ennis is late. Gannon will give him the angry face. And of course, as always, the men will eat it up. With the one gesture, his status will rise. It doesn't take effort to win respect, it requires only identification of the greatest common enemy, and then to become its most vocal critic.

Gannon, who fought with valor at Sigma Fugit, hated with passion.

Pryce, who killed a million innocent people at Amsterdam . . . and they're bloody loving it.

How is it this easy.

the descent

"Obviously we're here," Gannon tells the thousand and change men in one of Philadelphia's amphitheatres, "because we're damn good at what we do."

High up, where it's dark, Ennis drifts in and out. Gannon's voice reverberates across the PA, cracking not from stage fright or strain, but anticipation.

"It's easy to fool yourself into thinking that the enemy has some sort of advantage, that he may know something or possess a weapon that we simply don't. He may be quicker than us. Maybe smarter. Well?"

A pregnant pause. Even Ennis looks up, curious to see where this is going.

"He might."

Gannon looks up at the huge display behind him. Surveillance, mug shots, terrain. Not some jungle on Titan this time. Not a mountain range on Io.

"Mount Saint Helens is your first landmark coming in. You'll see it on the port side—a lot of smoke, lot of fire. Volcanic activity has ripped most of the North American continent apart, Yellowstone fizzled but that's the only place she got lucky. It's a mess down there across three faults, close to four billion have fled or perished. Expect to see bodies, a lot of bodies, everywhere.

Tsunamis are raging across the planet."

The displays behind him change to terrain maps, mostly Capital Expanse militarized regions, some Healthcom, EU based.

"Slingshot across the arctic on approach, we're low over Stockholm and Berlin, boots down on Roman soil nine minutes fifty seconds after atmo. Questions!"

A few hands go up near the front, middle management. Gannon points at one.

"Sir, with no disrespect . . . " Gannon nods. "Some are stuck on the notion that we're violating the Earth Accord."

"That's fair," says Gannon, "but we're Ombudsman-sanctioned on this one. No more friendly fire, no tribunals."

Ennis watches his eyes, waiting for the glance. It comes right away—the almost guilty, sly look that says they're the only two men in the room who know Gannon is lying, and neither will say a word until a good deal of Europe lies in ruins.

parley

Aboard one of three hundred gofast troop transports, encountering mild turbulence, Ennis can feel his PDA vibrate against the armor's breastplate.

"Personal," he tells it, touching a finger to his right earbud. "Hello?"

"Ennis, it's Apple Zenith."

He takes a sharp breath. "Yeah?"

"Is this a bad time?"

He looks around the red-lit interior, the seven soldiers who hope, if nothing else, to fire a rifle in anger this time.

"I'm good. What do you . . . ?"

"I'm no longer a part of the Amsterdam Trust."

"Okay."

"You're wondering why I would call to tell you."

"Yeah."

"I formed the trust as a means to an end. The past few years, I've kind of softened on that end."

"The killing of me."

"Right."

"Can I ask why?"

"Sure."

A pause. "Okay. Um . . . why?"

"It's actually something you said. On one of those Vanderbolts."

"Jeez, I was on acid for most of those. What did I say?"

"Something like, once you kill one guy, you might as well have killed a million. It's the same thing. It's murder. Small scale, grand scale . . . it's killing. I don't want to end up like you."

Ennis is quiet, looking at the men, wondering if they will see their commander break a little, here on the verge of what might come.

"Can I ask you something Ennis?"

"Of course."

"Does it still eat you up?"

He repeats the question internally, several times. "No."

"How long did it take?"

"A year. Then three. Then another three. Then six months . . . then about five solid years, and finally I said fuck it, I have regular access to pain management on the combat plan. From there, things have been reasonably . . . nice. Not ideal. Not great. But better. Sustainable."

"You wouldn't do it again? Kill innocent people."

"God no."

"I need to be certain, to put it behind us. I have your word?"

Again, Ennis looks at his troops. "In Amsterdam, there was a woman.

Her family were there for something like twenty generations, all very tight, traditional Dutch. When I had to leave I asked her to risk coming along, but she couldn't. She begged me to find another way . . . of course I needed the surgery, it wasn't optional . . . we both sensed the world falling apart around us but couldn't do anything to prevent it. I waded through paperwork for six years stuck in Rotterdam, close enough to touch the wall. As it became more obvious that I wasn't going to find a way in, she left me for someone local, sometime during the third year. I didn't get to hear about it until year five though."

Apple says, "Oh."

"There's more to the story than they tell you . . . but not much more. And not enough to say I couldn't have done any different. Point being, the situation won't ever come up again. She was the only thing in the universe that could make me so crazy, and now she's dust."

The cabin shudders, blue sky then horizon becoming visible through the portal while their transport slows to terminal. Far below, the fires of St. Helens roll a carpet of black ash across the northern tropopause.

Apple's voice grows distant, as if turning her head from the mic. "When people talk about it now, they call it a battle."

"Battle of Amsterdam?"

"Yeah."

"I overheard that once. I thought it was a joke, maybe for my benefit. I guess it almost would be funny . . . "

"Six minutes!" shouts the co-pilot over comms.

Ennis presses his brow to the cold glass, watching their descent into black. "I have to go soon. When I'm done with this whole war in Europe, can we meet?"

"Meet? Yeah. Maybe."

tactical retreats

A short time to catch the breath between barrages of GTG-009 artillery fire, and Ennis finally notices that he's bleeding.

He asks Lt. Mattias, "How far out is that drone?"

Watching it sweep around their position in a low roll, its ancient engine struggling even with simple maneuvers, Mattias replies, "Out of range."

"Awesome. Do we have orbital?"

"No. Wait . . . no."

Ennis shouts, "Can anyone here shoot around mountains?"

"We could probably wait until it comes around to launch at us again."

"That's a good call, Lieutenant. A grade-A Napoleon fucking Bonaparte move on your behalf. Does your combat prowess extend to treating a serious shrapnel wound?"

Mattias lowers the scope, looking at Ennis's leg. "That looks bad, sir."

"Cheers!"

mostly there

A squad of five—three original, two drafted from the remains of other gofast teams—approach the Corsican coast on foot with apprehension, all too aware that a good ninety per cent of the soldiers encountered on the causeway departed from the border towns ahead.

A shot rings out, landing nearby. Another, then like fireworks, a shower of sniper fire, muzzle flashes in the dark windows of skyscrapers half burned down.

This promises to be urban combat at its bleakest, an enemy familiar with the locale, and deeply committed to defending it. There's no convincing themselves otherwise.

Ennis guides them to cover, hoping bomb-damaged concrete and steel will be enough to let them wait it out.

"Maybe they'll run dry," says Mattias.

Ennis feels wet beneath the thigh. The bandage isn't holding.

While he adjusts it, Sergeant Grieves says, "I did what I could. I'm not a doctor."

Ennis frowns. "You're wearing a helmet with a big red cross on it."

"I took it off a corpse. Mine fell into the river."

Blake and Wentworth take the initiative, rolling out from behind either side of their cover, running to the overturned soda truck at the causeway's end.

An RPG hits the opposite side, shunting both men back into the open. Blake slams his back against the front tire while Wentworth takes their fire, dying quickly.

Blake leans around the wheel, firing off a clip at nothing.

"Hold your fire," yells Ennis, padding at his wound with a scarf. "Just stay there and wait for air support."

"What air support?" Blake's eyes fixed on Wentworth.

"Just hold there."

Another RPG spirals in, skimming past the truck, landing behind the others with a short, concussive pop, then the rainstick noise of falling debris.

As sand and shrapnel rain down Mattias yells, "RPG!"

Touching his left ear, shooting another syrette into the leg, Ennis says, "Call Gannon."

understanding

As it catches dying sunlight, Ennis can see the orbital rail track across a twilight sky.

"You fuckshitters have no idea," screams Blake. More sniper fire. Blake howls with laughter.

On audio, Gannon takes him off hold. "Ennis?"

"I'm here."

"You have the ball."

Ennis looks down at his PDA, flashing an incoming session. Accepting the prompt, it opens multiple command windows, each feeding realtime data on the rail.

He swipes to Corsica, zooming on the coast, their no-win-situation terrain instantly recognizable in realtime. Ennis double taps to highlight each of the charred buildings, sending a fire command.

Behind him, the strongholds implode, left to right in time with the scream of kinetic rods.

Ennis stands, then the others. The snipers no longer shoot at them.

"Gannon? We're good here."

"Ok. Keep the rail handy, Worldsnet air support was just retasked to us in South America."

Ennis pats Blake on the shoulder, pointing at smoking remnants. Blake nods, rounding up the other two.

"South America?"

"Yeah." Gannon sounds distracted.

"Is this it?"

"It?"

"Is it hitting the fan?"

"Yeah."

"If you had to guess . . . " Ennis checks the battery on his sight. "Why are Healthcom invading South America now?"

"Ombudsman's op districts are all there until March. Worldsnet are pulling out of Earth, it's official. Healthcom have boots on the ground already in Rio, they're trying to commandeer our server hardware before we can nuke the sites."

"Why is everyone leaving Earth?"

"They're setting me up at a command point on Faulkland-G. I'll no doubt talk to you soon."

"There's something bigger happening here."

"Worldsnet have promised us every resource required."

"Are you sending reinforcements?"

"How many of you left?"

"Three."

"Including you?"

"No, plus me."

"Well . . . you should be right. We don't anticipate heavy resistance beyond the border towns, and you have a rail. Good luck."

Ennis follows the others over rubble foothills.

mop and bucket

Everyone has now grown weary of shooting their weapons. Most of Corsica lies in ruins behind the four.

The city grid drops block by block ahead of their position, the smoke and dust giving them sufficient cover to avoid another loss.

"Three more," he says to the others. "Extraction's through the trees."

Blake, holding his ear, turns to Mattias. "They're taking us direct to Cape Horn. That's an eight minute breather." He looks at Ennis. "No one bothered to tell you?"

Ennis shrugs.

Their helo swoops in silent overhead, pitching a hasty landing beyond the skeletal tree line.

detour

Mike Hemmer watches them board until the last foot is in then punches it, not willing to risk another second at ground level.The TH-14 Huntsman lifts off first with a sluggish lunge, then a kick of power from the magdrive, pulling it quickly across the leveled island.

For Ennis, it brings back images of disaster scenes. The way Lisbon looked after half of the Canary Islands slid into the sea. The way Amsterdam looked. Again, not a memory of his own, only stock footage.

And if there's irony or some sort of satire to be found in the fact that this is the first destruction of his own making that he's actually witnessed . . . it remains lost on Ennis.

He walks from the cramped rear to the two man crew in front.

"We're not going to Cape Horn just yet."

The co-pilot turns to him. "What do you mean?"

"You're dropping me off first, in Singapore."

"I don't follow?"

"You don't need to."

The pilots look at each other.

"What about Gannon?"

"He'll wait. He owes me one or two."

awol

Weaver leans over the mushroom farm's lid, adjusting the aquarium tubing that snakes between mycelium jars, tiny fungi beginning to sprout within the folds of beehive-looking cakes.

Ennis sits on Weaver's couch, staring at Antarctica. He chews quickly, swallowing, taking another handful of half-dried mushrooms from the stone bowl.

"This is a live feed?"

"You asked me that already."

"Oh."

The sound of Ennis chewing fills the room as one meditation-conducive track winds down and the next ramps up to 'moderate'.

"It seems like every time I set foot on Earth something bad happens."

"Especially," says Weaver, pausing to blow into the tube, "when you arrive with an invading army."

"We didn't invade. We took back reclaimed territories."

Weaver replaces the small farm's cover, laying down on the floor. "Did you ever ask them why?"

"No. But I know anyway."

Weaver reaches into his cardigan pocket, shaking the yellowed glass bottle that Ennis brought. "Can you get more of this?"

"Sure."

dealbreaker

.Oporto Commercial.AT/CBD01—EU.2165.01.26.15:11

Apple strides through the elevator doors waving a PDA at Sylber.

"This is some *bull*shit."

"I'll call you back," says Sylber Giles, not looking up right away. "Apple . . . "

She tosses the PDA right at him, a direct hit to the bridge of the nose.

"Hey fucking dammit!" He swipes tissues from the desk's top drawer, dabbing blood.

"You can have the trust, Sylber."

"That's pretty much what's going to happen. Yes. Were you under the impression they're seeking your permission? I'm sorry, Apple. They should have . . . maybe made that clearer for you."

She leans across the desk, trying to slap him, but he steps back. "Whoa. That's hostile."

"This whole abortion of a revenge caper stopped being a force for good long enough ago that I no longer give a shit. Take it all. But God help you if I find your fingerprints on Tampa Valley. Because that's the day I start looking into you and yours. And I won't make the same mistakes."

"Apple, you're coming off like kind of a cunt right now. I think you should leave."

square one

Years of her life all but wasted. A billion or so in personal funds. She won't see it again, but fuck that. It's over, and she's clear. Legally, morally, financially. Whatever. Because there's always more where that came from.

So the proverbial what next.

Was she born to hold a grudge? Is her purpose tied forever to those who have wronged her? Maybe a constructive approach. How would someone in her position do some good in the worlds? And what is the exact opposite of everything she was before?

Absently scrolling through her address book for people to delete, one name in particular stands out.

She wants to laugh at the impulse—but by the time she understands it, Apple has already tapped 'Call'.

research

.Departitioned Uninhab.OM/DP4235—North America.2165.04.02.12:43

Nine months after leaving the trust, Apple leads the eighty man team from Valerian Geothermal into the valley floor south of Orlando Caldera, directing them to spread out in pairs, programming portable diggers.

Apple's PDA tracks the location of three unidentified surveillance drones that maintain a somewhat stealthy twelve-click holding pattern around the former disaster area.

She holds back the urge to flip him off, secure in the knowledge that her very being here is enough to have Sylber shitting his designer slacks.

secret meeting

Napping in the cabin of her private transport, Apple is woken by the driver turning back to her. "I think this is him."

She looks out through the window at a rental cruiser, parked in the shade of a tall tree. Sitting on its exit stairs is Ennis Pryce, his cautious gaze fixed on her window.

For whatever reason she feels compelled to wave, and he nods back.

---

parallels

.Montana Shareholder.LM/ST14—North America.2169.02.15.17:00

“I knew as soon as you said it, that the ‘two planets colliding’ were Otis and me. The two people who would tear each other’s lives apart. It was obvious even then. But I went ahead and did it. And I found out the things I did. All of the bad stuff happened. He left, I moved, the smoke cleared . . . and then I thought, so wait, that was shitty, but not profoundly devastating. I thought it was supposed to be, you know, planets colliding.”

The fortune teller chews, something crunchy.

“So for the longest time, I thought you were wrong,” says Imogen. “I started thinking about what happened, gaining some perspective. Maybe losing it, I don’t know. I can’t sleep now. I literally *can’t* sleep. I’m finally getting things done. Stringing thoughts together. Work. I haven’t touched caffeine or pep in six months. It’s horrible, and great. I’m a person, but not a real one.”

On audio, the fortune teller says, “You called him?”

“Yes.”

“Mm.”

“Things go bad if we’re together again, right?”

Silence, then, “Yes.”

“That’s what you were warning me about. Not last time, it was this time.”

“Yes.”

Imogen sighs. “Can I do anything about it? Change the outcome?”

“Of course!”

"And can you tell me what it is?"

"That falls into the same category as lottery numbers."

"Just this one thing then. Is there any way for me to be with Otis, and be happy?"

"Yes."

"And then . . . will *he* be happy?"

"No."

Eight

trial arrangement

.Glasgow Residential.HM/HP01—EU.2169.06.01.11:03

Ennis juggles the PDA casually, waiting to disembark.

For a moment he wonders if the item is his only leverage, a brief flutter of fear that's hard to really sustain. On the stairway he staggers, knees loose, thighs twitching. He sits a moment, patting himself down for the joint he's ninety per cent sure he brought.

Apple's Bentley is pulling through the park's main gates, slowly onto the landscaped lawn.

If she wanted to kill him, this would be the perfect chance. A sniper in the playground's crow's nest . . . an exploding Bentley . . . with its tinted windows he can't even see if she's in there.

But she's had her chances. If she turns out to be the one person he can take at face value, he would need to rethink his stance on irony.

A driver exits the transport, opening its rear door. She waits in there, tired-looking.

"You can't smoke that in here."

"That's okay." Ennis takes quick drags to light the tip, straining to hold the first lungful. "I'm good here for a bit."

She gets out, patting down her skirt before joining him at the stairs.

Ennis offers her the weed, which she takes, puffs, then hits again.

"This is surreal, right?" The THC potentiates an earlier opiate dose, settling post-flight nerves, breaking the tension to some degree. She seems to sense it too, and nods.

"A little."

designs by committee

.Bailey Restricted.OM/SS3455—Secure Space.2169.06.08

General Gannon enters the minister's office with a sharp salute, sitting opposite the older, smaller man.

"I forget," says Minister Gomez, "you had an uncle who made General?"

"Grandfather."

"Five stars, no?"

"That's correct."

"What was he famous for, the original Gannon? It's on the tip of my tongue."

Gannon shifts, imperceptible but to him. "The stockade."

"Of course." Gomez peers at him over glowing lenses. "Forgive me for asking, but our side or theirs?"

"Ours, sir."

"Of course." On the desk, Gomez's PDA vibrates. "That's Emmerton."

"No doubt you're already aware, but he'll be calling you to advise that we just lost the western districts of Feedcast."

"I am aware . . . but it's okay. We factored it in as acceptable."

"They have a major feed network now. It could cause diplomatic fallout if they start broadcasting their theory."

Gomez nods. "Which brings us back to Pryce."

It hangs in the air a moment longer, each seeming to want the other to go first.

Finally Gannon says, "We heard he's linked up with Apple Zenith and her new deal."

"And what about this other rumor?"

"It's true as well. He was presented with everything that exists on Tampa Valley. Best guess, she now has it too."

wildcard

.Darwin Residental.HB/FG782—Australia.2169.06.14.22:13

Eli Lylo wraps a gloved hand gently around Hugh Po's mouth, the latter aware of the act only when his head is jerked back, ear held to Eli's whispering lips.

"The pinch you feel beneath your right shoulder blade is my diving knife, entering between the fifth and sixth rib."

Hugh doesn't make a sound.

"It's not fatal at this depth."

Hugh seems to get it all at once, slowly nodding, murmuring something through the hand.

Eli pulls it away. "Say it again?"

"I know what you're here for. I know where they are. Please take it out. Just take it out."

"Okay."

"Please."

Eli removes the knife with surgical care.

"Apple's boarding a flight to Havana right now. Pryce is already there."

Eli wipes his knife clean. "What's her angle?"

Hugh shrugs, wincing at the pain. "I assume she wants to get back at him somehow. For Amsterdam, her mom died there."

"Hm."

They lock eyes a moment.

"Don't kill me. Please? I'm worried that you're still going to kill me even though I told you all that stuff."

Eli smiles. "That would be a dumbshit move on my part. You're my man on the inside now, Hugh."

"Okay. I'm cool with that."

things to do in Cuba

.Havana Entertainment.GB/B65—North America.2169.06.15.19:19

The first hint of something awry that Ennis senses is the look a bartender flashes him on the busy corner of Goro and Fez avenues.

He's seen that expression once. Where was it?

Tightly packed streets give way to a quieter area, a former industrial neighborhood reclaimed as employee boardwalks lining Havana's southernmost shores.

The second clue to what awaits is a brand new Lexus parked opposite Apple's hotel, a standout on the island of smartcars and mopeds.

Third, and foremost, he recognizes a man in the crowd as Eli Lylo, following at a distance. That's definitely an indicator of bad tidings.

exodus

.Victoria Residential.QB/EH808—Triton.2173.08.01.08:56

Umberto checks the duffel's zippers again, patting down his pockets.

Just beyond the bedroom door he hears Monique pause in the same place as always. She won't say it today. She doesn't have the nerve.

Just go, he wills her. Get your keys and leave. Soon he hears the jingle, his ears tracing her exit through the apartment door.

This is the fourth place they've lived on Triton, and the last.

Her car starts with a hum, stuttering before making its way through the car park gates. He lives for that sound, today more than any.

Umberto whips the front door open, looking around before stepping outside, locking it behind him. A cold morning. He wishes he packed a pair of gloves, but not enough to go back in.

joyride

At Medevac Asiatic's small Triton airfield, Umberto finds no one at the front gate, almost disappointed that his flawless cover story will go to waste. He steps off the moped, leaving it running while he leans into the guard post, opening the boom himself.

Pulling up by a pair of hangars on the northernmost end of the facility, he finds them similarly deserted. While prepping the spooler on a UH-88 Jarhead heavy-lifting helo, Umberto spots a lone mechanic wandering aimlessly into the hangar, visibly perturbed.

"Hey!" snaps Umberto, conscious of the slight quiver to his voice. "Are you supposed to be in here?"

The mechanic glances up, eyes red.

"They shot all our guys down," sobs the mechanic. "Jesus man, they were squawking medevac."

Umberto looks around the empty hangar again. The deserted landing pads beyond its open doors.

"Where were they going?"

"Io."

"They're all dead?"

"It was a trick. It was set up by one of the other two."

"Capital?"

He shrugs, then gradually becomes aware of Umberto's activities. "Wait. Where are you going?"

"Io. I'm on recon for the directors."

"I didn't see anything come through about it."

"And you won't. No one will."

The mechanic pats himself down for a cigarette, sitting on the floor. "There's no point. We saw them all drop out of the sky."

Umberto begins disconnecting the fuel hose, polling the old hauler's avionics with his free hand.

onset

.Victoria Transit.OM/TR50—Triton.2173.08.01.12:04

Monique sits in Caroline's Mercedes watching floodlit rain pound cement beyond the Laughie's carpark overhang.

She waits quietly until Caroline returns to the driver's seat with snacks and drinks, handing them to Monique before pulling back into traffic.

Finally Caroline says, "Okay keep going."

"Well he acts like I encourage these guys."

"That's no good."

"He thinks I go home with them." Monique throws up her hands. "What am I supposed to tell him?"

"Maybe you're reading too much into it."

As they come around the tight bend Monique is first to spot the overturned car.

Caroline pulls over, both of them dashing from the Merc, through the destroyed crash barrier to the upside-down driver, hanging by his seatbelt.

"He's breathing," says Caroline, taking a moment to study the blue spiderweb pattern across the man's face and neck. "Wait . . ."

"I know this guy," says Monique. "His kid is in my class."

"He's showing symptoms."

"What?"

Caroline stands, navigating emergency menus on her PDA. "We both need to be in quarantine. Him too."

Monique stares at his face. "This isn't . . . the same thing?"

"I don't know."

"You said I'm not infectious."

"You weren't. You shouldn't be at this stage. I don't know okay?"

vacation

.Aukland Nature Reserve.HO/EG5106—Triton.2173.08.01.16:27

Recently, maybe since the flight from his apartment, the places that depress him seem to inspire Umberto. He knows that's why he was drawn to the midtown rail loop. Possibly, it's why he was drawn to Jess. And of course Monique.

Climbing the small foothills beneath a cliff that dwarfs all around it, he begins catching glimpses of the beach house at its edge.

A stairway cut into the rock zig-zags all the way up from here. It looms ominous, but possible. He rests three times.

At the cliff's edge he stands sweating in the front yard of an improbably blocky mansion, all tinted glass and steel edges, shimmering under midday sun.

At the front door he waits, staring at the sensor while its display fails to appear.

"Doorbell", he says again.

Eventually Jess appears at the door, unable to muster a smile at what even he perceives as an intrusion.

"I know you told me it's not gonna happen . . . " he begins.

"Why did you come?"

"I'm leaving now. I have a helo parked on the beach. It's up to you if you want to come."

"You know I can't."

"I can get you off-world without a problem, and around Quarantine on most moons."

She looks back inside, eyes darkened by the home's interior gloom.

"No. I can't leave him alone."

Umberto fails to find a followup, electing instead to simply nod, back away slowly, and leave the way he came.

Jess watches him disappear beyond the steps of her father's great stone wall, again hoping that he glances back.

reprieve

When he's high enough that the sand begins to settle, Umberto brings the Jarhead's nose around to face the cliffs, his eye caught by movement on the rocks above.

A lone figure descends the stairs three at a time, barely visible.

He brings up an optics panel, zooming in on Jess.

The big helo lumbers, rocking to port, starboard, before stabilizing, nose aimed at the cliff face.

Jess freezes, raising a single hand to him, waving it slowly.

Moving in closer on her face, Umberto can see that she's crying.

Weaver's wish

.Havana Accomodation.MS/34500—North America.2169.06.15.21:07

Ennis enters the hotel lobby aware of three hostiles in his peripheral, controlling with intense focus the neck twitches that would give his intent away. They can't know he knows. He needs another thirty-one seconds.

In the relatively subdued, marble lined lobby, Ennis's suddenly ringing PDA is like dropping cutlery.

He answers it the same way he imagines he might normally answer a call, although now that he's thinking about it, the notion is completely alien to him. How does he normally answer the PDA?

"Hello Ennis speaking, this is Ennis . . . "

"Is that . . . Ennis?"

"Weaver?"

On audio, Weaver says, "You're not a solid, yeah? You're not even a liquid. Okay?"

"What can I do for you?"

Weaver croaks, "Is now a bad time?"

Ennis hesitates. "Not at all. Been doing some acid there bud?"

"DMT . . . God. We're gods. I feel emblazoned. Is that right? Emblazered? Bedazzled."

Ennis trots upstairs to the mezzanine bar, edging between a light evening crowd.

"Weaver I'm sending something to you now, I don't want you to try and read it, just save the file and take a look when you're feeling more . . . you know."

He pulls the PDA away from his ear to batch an upload before chancing a look back at Eli, also pushing though the crowd.

"Hey Ennis, what was the best thing you ever did?"

"Huh?"

"Like the best . . . time in your life? When was it?"

"I don't know."

"Okay, then think about it. Think really hard. Put it in your mind and keep it there . . . then next time I see you, I'll show you how to go back. I will *literally* send you back in time. Well not literally."

Ennis feels his throat tighten. "You can do that?"

"Yeah man. I synthesized something. It's not hallucinogenic, not dissociative. Just something else."

He enters a long corridor that leads to the kitchen and bathrooms, turning immediately left into staff change rooms. "Weaver I gotta go. I'll come see you Monday." Through the far door and into an alley that runs the length of the building, Ennis turns right, back to the street. Coming up behind the idling Lexus, face turned away from its mirrors, he slides into the back seat, resting a small pistol on the driver's shoulder.

"Hi," says Ennis. "I doubt they'll hold their fire just because you're in here with me. You should probably go go go."

The driver starts it up.

the second time

.Miami Transit.BA/SW4467—North America.2167.09.02.07:14

Apple answers her PDA still chewing, taking water from her bag. "Ennis."

"Eli is here. He knows where your hotel is. I have one of his goons with me now. I'm holding a gun on him, making him look like a total little bitch. He can hear me, he's just not choosing to acknowledge it. Jake. Jake! Come on Jake."

In the background, "My name is John you dick."

Apple asks, "What do we do?"

"How far out is your flight?"

"Ten minutes."

"You know the Amber Lounge in Havana-C?"

"Yeah."

The line clicks dead. As she puts the PDA down, it rings again.

"What?"

"Apple it's Monique."

Apple says, "Oh my God. Hi . . . "

"Greg's dead."

"What?"

A silence that Apple soon realizes is stifled sobbing.

"Monique what happened?"

"They have Liam and Deb. I'm being coerced . . . to board a ship. They told me not to talk to anyone, but I have to. Sorry, it was you or dad."

"What? It's okay just tell me, what happened to Greg?"

"They set it up to look like it was my fault. Injected him with some . . . virus. I don't know. They have my kids, Apple . . . I'm driving to the airport, they told me to board AJL-7050 at midnight. It's not a commercial flight."

"Monique . . . "

"I don't know what they're gonna do."

"Don't get on the ship."

"They have the *kids*."

"I'll help you. We can get them back."

"Good. Whatever happens, please do that."

"Wait—"

axis

Both Ennis and John Weng jump as Apple throws the door open, slamming it against the wall.

She steps around Ennis to place a snub-nosed revolver against John's dick, tugging back the hammer. Ennis chews a Morpheze, watching on.

"Hey. Guy. Did Sylber Giles split the crust open at Tampa Valley?"

"Yeah. It was Sylber."

"Jesus Christ Ennis, why didn't we just do this in the first place?" She rests the muzzle on his chin, scratching it. "Did he use that orbital 'sensor array'? Is it a weapon?"

"Yeah. Well no, there were charges onsite. But he has a weapon there on the moon. No idea what it does."

"How *does* it work then?" Ennis interjects.

"What did I just say?"

"Sorry."

Apple lowers the gun, having made her point. "Why did he do it?"

Weng shrugs. "He just told us to set it up. Didn't say why."

"John, did Sylber arrange for my sister's kids to be kidnapped today?"

John looks up at her, puzzled. "I don't think he even knows you have a sister. I didn't."

somebody's data

Weaver's call comes a few minutes after Ennis has dozed off, the hooting and thudding bass still bleeding through from three floors below.

"How you feeling champ?"

"That stuff you sent," says Weaver, "what's the deal?"

"Ah it's all good man, I thought shit was gonna go down with this guy who was chasing me but I totally turned the tables on his crew. Just hold onto it in case we go missing here in Cuba."

"You wanted me to decrypt it right? It's okay if I look at this?"

"Nah I tried. I lined up a guy but he's being all coy about it."

"I opened the files, they're in your q-mail. I read all of it sorry."

"I thought you were a physicist slash chemist dude, I didn't even think to ask."

"There's some serious shit in there. The Amsterdam Trust did Tampa Valley."

"Yeah I heard something along those lines. Is it proof? Would it hold up in court?"

Weaver laughs. "What court? Ombudsman?"

"Hm. Point."

"There's plenty of other stuff mentioned. Probably something that could shut them down if you use it right. Come hang out sometime if you want, I made notes but a lot of it's still rattling around."

"I'll be done here Sunday night. We'll come straight to you."

"I don't know if you should bring Apple Zenith. This was all going on around the time she left, but some of it runs further back."

"Apple's cool," says Ennis. "But what do you know about Sylber Giles?"

"He's a total fag."

"Huh."

"Yeah he's gay. Really gay. He enjoys a sack lunch if you will. Gets his daily serving of—"

"Do you think we could send a hot man assassin to his hotel or something?"

A rustling on Weaver's end, then the unmistakable sound of a bong hit.

"Definitely. Probably. I dunno. He's queer, not stupid."

"Do you hate gay guys or something?"

"Yeah I guess. This little gay went to university with us, when my wife and I were first dating he told her he wanted to turn straight for her. Tried to break us up. It probably stuck with me."

Ennis is quiet, half asleep, aware of something tragic in those words although unable to feel anything but mentally bloated.

"I'm gonna sleep. I'll call you when we hit atmo."

"Hey Ennis, before you go . . . there's something in here about a giant planet."

"Tyche? Marduk?"

"Yeah. Is it true?"

"That's the big question."

"Well why don't we go out there? Take a look."

"No one knows where it is."

Ennis can hear that Weaver is smiling when he says, "The owner of your encrypted PDA knew."

encrypted pda > intercept > tampa.valley

>The group is willing to proceed. Please note that emergency services will be diverted as well as communications disrupted locally to ensure efficiency on the day. As we anticipate some civilian disorder, see that your exit strategy is concealed and secure in order to guarantee the team's immediate withdrawal from operational areas.

leaving the island

Apple and Ennis sit alone in the hotel's rooftop bar on a Sunday, looking over a series of building plans on his PDA.

"This whole thing is controlled from the surface of Luna", he tells her, bringing up a territorial map. "Giles has been snatching up any piece of real estate east of the Fritz crater system since before you left the trust."

"Could she have faked everything on the PDA?"

"Probably. But pretty unlikely. What's the motive?"

Apple bites into a vegan quiche, inhaling sharply at the heat. "We could probably confront Sylber with it. Get a confession on record?"

"What about Lylo?"

"I don't know." She breaks up the rest of it with her fork, watching it steam. "I had a good relationship with the guy. If it's business, we could outbid Giles. If it's politics, maybe sway Eli with what's on the PDA."

"You know what? Send it to him."

"Which bit?"

"All of it. Send it to Eli, he goes to confirm it with Sylber. That conversation could be the catalyst for a split, probably not, but either way the threat is conveyed passively. Sylber will have no idea what we want, or plan to do. Just that we have it, and are eager to share."

Apple picks up her PDA, dumping the entire contents of their recently-decrypted PDA into a q-mail addressed to Eli Lylo.

"Is there a downside?"

"Plenty," says Ennis as she hits send. "Especially if Sylber was just acting as proxy for Goseph Mailer."

conversion rate

"You people need to calm the fuck down," says Eli Lylo on audio, voice distorting, "meet me at the airport in twenty. And Ennis? Bring my driver back."

Apple leans over the PDA. "Eli? Did you help Sylber set up Tampa?"

"Apple. You're better than this. Meet me at the airport and we'll talk shop. Gate TH-009."

white rum

Havana's only departure lounge is dead but for Ennis and Eli, sitting a seat apart under the glowing Bacardi sign. Way out on the tarmac, a Venus-bound gas miner lights its engines, the glass walls and bottles reverberating.

Ennis is cocking his head at Eli, mid-taken-aback. "Everything?"

"Yeah. Sylber's Ombudsman, Ennis. He's their guy on the inside. Apple didn't know what to look for back then, but I'm surprised she hasn't figured it out now . . . they just gave him a compelling back story and sent him in for the CFO role."

"What's your deal then? How do you know this?"

"I'm cointel with a different org. The trust's sponsorship was my cover, we were watching Giles since he entered the picture."

"You weren't gonna kill me the other day?"

"I was going to sit you down and tell you that we're on the same side."

"Well."

"I was also going to brief you on the trust's new Lunar operation."

"We're nut deep in it, Apple wants to shut it down."

"We do too," says Eli, "and I'm thinking you might be able to assist."

Ennis sighs. "Don't say it okay? The thought crossed my mind, but no."

"This time we can get all the people out. We start with his infrastructure there, and we take out the whole damn command hub."

"Get all the people out? It's like thirty square clicks."

"Of automated rigs and security. Maintained by an eleven man crew."

"How do we get the crew out?" asks Ennis.

"First, how do we get hold of your guy with the nukes?"

"You're some kind of super spy, can't you sort this out yourself?"

"I'm out here all alone. The way we operate is to manipulate chumps like you into doing these things for us."

Ennis shrugs. "Fuck it. You want what I want. Let's talk to Otis."

born again

.Rotterdam Transit.OM/UH45—EU.2169.01.29.22:48

From the idling rental, Otis watches Keo's safehouse with starlight lenses, having spent the last three hours waiting for any sign of occupancy.

Otis is pretty sure he's in there. Probably down in the bomb shelter. It's the only place they wouldn't know about yet.

A crackhead meanders by, scoping out his ride, unable to see Otis eyeing him back. His hand instinctively drops to a sidearm, but relaxes as the guy continues further into the neighborhood.

Finally Otis gets out, arming the car's security while walking to Keo's side gate.

timing is everything

A girl of about twenty is exiting the shelter when Otis enters the backyard, giving her a wide berth. He catches the hatch before it hisses closed all the way, patting himself down before descending the ladder.

Sprawled on the pink couch, fatter than Otis has ever seen him, Keo watches the feed with a beer to his lips.

"Keo." Otis raises his sidearm, the sites locking in on Keo as he rolls to swollen feet, arms raised.

"Otis what the fuck, man? Where have you been?"

"You're quick for a guy of your size. Is it true?"

"Go on."

"Ombudsman had you informing on me?"

"Yeah."

"You're a snake. A fucking . . . "

"Hear me out."

"Okay."

Keo blinks. "Um."

"That's about what I thought."

"They already knew how close we were, bro. If I didn't play ball they would have disappeared us."

"You should have told me what was going on dude."

Keo nods, looking at his feet. "I know . . . "

Otis lowers the sidearm. "No more of this shit okay? Let me in on everything. Partners right?"

"Sure. Sorry."

"Listen I need a nuke."

Keo takes a moment to breathe, placing his beer on the coffee table. "Hmmm. You kind of ballsed up on that whole deal last time. Looked bad for both of us."

"Hey, that was your client originally."

"So who's it for?"

"Same guy."

A pause. "Right. If I get it for you, forgive and forget?" Keo's eyes never leave the weapon at Otis's side.

"It's forgotten. We're cool. Somewhere round forty-five megaton."

"Okay, in about a week. Beer?"

"Nah. I don't drink."

As far as Otis can tell, Keo seems to be buying his transition from homicidal to easygoing.

How fucking naïve Keo must believe him to be. But maybe Keo doesn't believe shit. In which case, Otis truly is naïve, and will only really know it when he feels the barrel pressed somewhere just outside of peripheral.

filler

.Biway Triton-Earth.OMHB57359—Transit Space.2173.08.04.12:21

Without any prompting from Umberto, after hours of uninterrupted silence, Jess just launches into it.

"The alliance wanted to go from excluding unpleasant events in the feedcast to rewriting them entirely. My dad tried to kick off an enquiry through Ombudsman and was arrested."

Hyperspace biway gates flash by every few seconds, flooding the cockpit briefly white.

"Diplomats don't serve time in prison, they're just placed under house arrest and firewalled, so they won't bother anyone."

She's quiet a while, and then, "At least he didn't back down."

Umberto feels an unspoken pressure to enter the conversation, grasping for an angle of approach.

"Is he going to be alright there by himself?"

"He's okay now."

"You got a nurse, or . . . ?"

"Dead."

"Oh no, wow," says Umberto, biting his lip. "I'm sorry, I didn't realize."

She watches his eyebrows descend, sees him piecing it together.

"So he was already gone . . . when I spoke to you at the house?"

"No."

Again his frown grows deeper. "Um . . . "

"Yeah. I gave him a lethal dose of Benzo and let him go in his sleep."

"What?"

She hardens, sitting up straighter in the boost seat. "There was no one else to do it. No one who would take care of him either way. And because he asked me to, for years before you came."

Umberto flicks a set of redundant toggles, aimlessly checking nav aids, still unsure of the right thing to say or do.

Finally, "Are you okay?"

Jess says, "It went . . . okay."

Facing his side window, Umberto mouths, "Yikes."

empty nest

.Near Earth Orbit.OM/EC419—Asia.2173.08.08.21:54

Requesting clearance again as they enter the stratosphere, Umberto is noticing the complete lack of traffic on Earth's skyways.

"Something's wrong. They're not responding."

Jess looks around outside, then at qdar. "Where is everyone?"

The Jarhead winds a lazy spiral down through the cloud layer as land masses grow larger below, vaguely unfamiliar. Coming in lower, it's evident to Umberto that the coastline has been redefined somehow.

Circling Mount Fuji, they're both looking for any sign of Tokyo.

Long range scans reveal most of mainland Asia similarly rearranged, inundated by flood, the water only recently having begun to recede. The islands, he notes, are all gone.

Central Australia provides at least a place to set down, regroup and perform wide area surveys.

Neither can really grasp it. Jess leans forward watching the churning sea, while Umberto reads text aloud from an Ombudsman emergency ticker, describing one of history's great disasters unfolding literally moments before their arrival.

Umberto scans for a feed network, but flicks it off almost instantly, knowing that every Medevac Asiatic vehicle he has previously piloted had any form of infotainment locked out by group policy.

silence reborn

Umberto and Jess traverse the empty corridors of Medevac's Australian Continental Logistics building. Dust has begun to collect on its surfaces. Coffee mugs sit abandoned on desks, spilling over with fuzzy mold.

"Did they know about the tsunamis in advance? Why else would they all leave?"

"I heard about an alliance war on Io . . . " Jess tries swiping her PDA past a feed station but finds it powerless. "Do you think it would have spilled over?"

"Not on Earth."

"There's gotta be—"

"Shh!" Umberto walks to the window, head tilted as he listens. The unmistakable whine of a twin-engine pleasure cruiser grows louder until the small Bosworth is visible weaving between Sky City Nine and Pepsi's twin mausoleum towers.

Umberto and Jess dash for the exit, downstairs and back into late afternoon sun, watching it land beside the much larger Jarhead.

"These guys must know something," he says to Jess. She hangs back a little, cautious, as two men exit the cargo door, both with assault rifles.

Umberto waves. "Hey!" The gunmen regard him with detached curiosity. "It's okay . . . we're friendly. I'm a medic, she's Jess."

"We followed you in from orbit," says the first, "where are you coming from?"

"Triton. What happened to Asia? Where's everyone gone?"

A woman also disembarks the Bosworth, something recognizable to Umberto in her expression, her manner. He might have seen her on the feed or in a movie.

She asks, "They don't feed news to Triton anymore?"

"Things are held pretty close to the chest there it seems."

"Our networks are heavily sanitized," adds Jess, verging on defensive at the implication that they should somehow know what has happened.

The second stranger says, "We need to borrow your helo."

Umberto takes a long, cautious breath. "Can I ask why?"

The men look at each other. "There's something heavy that we need to move."

"I guess it's yours either way. But before you take it, would you mind telling us what happened to all the people?"

"Everyone left," says the first man. "Earth's a dead planet now."

Umberto shakes his head, having seen it himself, still trying to find a path of logic to disprove the idea. "How?"

The woman points at the horizon, Umberto following her outstretched finger to sunset, and the new heavenly body straddling Sol.

The bearded guy tells him, "Gravity is about to swing it right back at us. Round two will be worse."

"We figure that's going to blow the Pacific caldera entirely," she adds. "Thc moon could even take a hit from its debris field."

The other man tells them, "You guys should probably come with me in the Bosworth. We're headed for the Black Sea. Our friend is asleep in back, he's somewhat older than the rest of us so try not to wake him please."

Umberto looks to Jess, then back. "I've been away . . . four years? Six maybe. How does this happen?"

"It's been coming a long time," the woman says. "They just chose not to tell us until they'd sucked the world dry and pulled out entirely."

Looking at her a long moment, forgetting his manners, Umberto says, "This is bothering me. I know you from somewhere."

"Apple Zenith," she says, "like the insurance guy."

say goodbye

.Madrid Shareholder.OM/SH6775—EU.2169.06.24.14:09

Otis carefully runs his tongue along the paper's seam, spitting a piece of finely-chopped cannabis onto the table while he seals it. "No. It's genetic. Every generation thinks it'll be the one to conquer death, so . . . "

Handing it to Imogen, he lights a match, shielding it from a warm wind that gusts in off the equatorial sea.

She lights the tip, taking her time to get it started.

As the tide comes in higher, waves crash louder on the beach below. Everything hits her at once. The smoke, the pink noise, the red tea.

Imogen dozes a moment, blinking awake, squinting when she turns to Otis, afternoon sun bathing her orange. "Do you believe in psychics?"

"Like Chip Geller?"

"Who?"

"The PsyFeed guy. With the costume?" He finally takes the joint from her.

"Oh yeah, I guess. More like the crazy old ladies you go see at flea markets."

"I'm not a big believer. I'm sure there's stuff out there we can't see, and there's probably ghosts and junk floating around . . . telepathic contact on some level, but someone who can talk to the other side and tune out the static? Predict the future?"

"Yeah."

"Don't buy it."

"Why?"

"There's too much going on," says Otis, "we're making too many choices every day. The universe can barely keep up, let alone flea market lady. I know it speaks to the whole free will versus fate shemozzle, but . . . you've heard Stairwell right?"

"What?"

"By Animal Magnet, it came out about twenty years ago . . . oh man, it just starts simple, then kind of builds, and unfolds . . . hang on." Otis unpockets his PDA, telling it, "Play Stairwell by Animal Magnet."

A soft but quickening harmony rolls across the balcony, meeting the waves.

"I had my first kiss to this song," says Otis, leaning back.

Imogen frowns. Behind the sunglasses, aware of where she's already gone, Otis frowns too.

"We met, what, sixteen years ago?"

"Right."

"How old are you Otis?"

"Okay, yeah. I lied about that too. Are you mad?"

"Not even."

"Really?"

"Yeah. It stopped affecting me. I suppose it's either acceptance, or that I no longer feel that particular frequency. Hah."

Otis kind of nods.

"So how old?"

"I'm thirty-four."

"Not forty-six."

"I never used life extension. This, believe it or not, is the face of an actual thirty-four year old."

"You told me once that people used to always treat you like a kid."

Otis nods.

"You wanted them to take you more seriously."

"They did . . . the moment I started equipping them. Junior high was pretty tense because of the arms race and all, but no one ever shot the guy with the small weapons locker."

She almost offers a smile. "I never even know when you're lying, or telling the truth, or even joking."

"That's all a part of the overall charm," says Otis, kissing her shoulder.

contact point

.Madrid Health.OM/SHH0001—EU.2169.06.25.10:40

"Where are you right now?" asks Ennis, audio only.

"Taking my girlfriend to the doctor."

"Oh. Sorry. Is she okay?"

Otis stands at the eightieth floor glass box balcony watching the maglev speed by between his feet.

"She has chronic fatigue. It went away for a while, but lately it's been a bit of an issue."

"Shit." A pause. "Can we talk about the thing?"

"Yeah. It's on the way. About a week."

"Cool."

Apprehension on Ennis's side. "So you're on Earth?"

"Yeah."

"Do you want to get a beer?"

Otis turns from the window, hesitant. "Not really . . . I don't drink."

"Movie?"

"You want to see a movie together?"

"Sure. When everything's good with your lady."

"Can I ask why?"

"I'm bored. I don't really know anyone on Earth now . . . just Apple, but we don't hang out. It's still weird when we're in a room together."

"Because—"

"Yeah."

"I need to buy some shoes tomorrow," says Otis at last, noticing Imogen return to the receptionist's desk, swiping her PDA. "Feel free to tag along. We can catch up, shoot the shit. Whatever."

Ending the call, Otis takes her hand. "Everything okay now?"

She nods, shrugs a bit, nods more. Something about her is different.

"What?" he prods. "What's that look?"

come clean

.Madrid Transit.OM/UH5—EU.2169.06.25.15:27

Driving home, waiting to enter the clearway streets, Otis can't help stealing glances at Imogen, trying to assess her emotional state.

"Remember I told you about the Amsterdam guy?"

"Prince—Pryce? I heard he works for Ombudsman now."

"No, he was involved with Worldsnet for a while. He got in touch with me again. He's caught up in a caper with Apple Zenith, trying to bring down her old company."

Entering Lauchlap Michael territory, Otis tilts his side mirror skyward, watching the drone that has been tracing their route through Ontario's urban hub from about two-thousand feet.

"He called you?"

Otis nods.

"Why?"

"This is all going to sound a bit weird."

She braces herself, aware only that anything could come next.

"A few years back two of the big three alliances went to war over mining rights on some of the outer moons."

"Ah, yeah, we were involved. A lot of the tactical briefs trickled down to us in the planning stages of Io and Europa."

Otis nods. "Do you know when the war ended?"

"Three and a half years ago. Worldsnet won."

"The thing is, it was never about mining, and it never really ended. Ombudsman and the big three have a secret . . . an untapped resource that every corp is going to want a piece of the moment it goes public. It's like, a big planetary system. A dwarf sun. Coming towards Sol."

Her blank stare.

"Yeah, that's everyone's reaction. This is going to be a bit shocking to hear but a huge planet or most likely star called Tyche—"

She laughs, cutting him off. Imogen seems to relax, shaking her head a bit. "Oh man, Otis . . . I thought it was something serious, you had me worried."

"What do you mean?"

"Someone's pulling your leg, Tyche is pseudoscience. They proved it doesn't exist."

"Right, but then they saw it appear in the sky beside Betelgeuse, and had to reevaluate the evidence."

"Are you trying to fuck with me or something?"

"No. Not at all. I think that's why everyone's been going so apeshit these past few years, Worldsnet and Healthcom going to war over every pissant piece of land from here to Sedna, McPherson sending me and Keo after Behemoth. Ombudsman has been falling out with some of their oldest allies over what everyone figured was some kind of huge secret. Tyche is coming. It's pretty much going to knock Earth into the sun, if their math is right. We have about four years left before it starts to affect the weather--another year until the quakes begin."

She just stifles a grin, still playful. "Why haven't they warned anyone huh?"

"Because right now they're dividing up the rest of the solar system amongst themselves while keeping core earth-based economies flowing. They want as smooth a transition as possible. I guarantee that they will wait until the last possible moment to feedcast this. Worldsnet already tried . . . Healthcom and Ombudsman stepped in, executed their board. Disappeared everyone around the leak."

"Gosh."

"A guy named Sylber Giles came into the Amsterdam Trust working for Ombudsman on the sly, they recruited him running low-level provocateur ops at protests he had helped organize. Apple had no idea. He eventually forced her out and graduated to false flags. He destroyed Tampa Valley, uncapped the volcanic caldera and flooded the canyon system on their behalf."

Imogen shakes her head. "Tampa Valley was a natural disaster. They said on the feed it should have erupted sooner."

"No. Sylber made it happen with explosives, maybe underground nukes or AMs. Ennis got a hold of his old PDA. It's all in there I guess. The problem is, there's no one to show it to. Ombudsman and the alliances all know the truth already. They'll do anything to seal it off."

"Why would they fake a natural disaster?"

"No idea. As cover? A way in, potentially.""

"How's that?"

"No military faction can operate here under the Earth Accord. After Tampa a call went out for aid, Healthcom responded with four million troops for rebuilding. Only they didn't rebuild . . . most of them took up fortified positions around alliance territories, and began evacuating their shareholders. Worldsnet came in with SEAL teams and a low-orbit fleet to lock down all their own resources and begin extracting assets to Mars and Triton... everything you've been seeing on the feed is a lie. Half this planet is already burning, and they're all just getting the hell out of here before we find out. The war was mainly about covering their exit I think."

"What good would that do them?"

"Because if it all collapses into anarchy before they get everything of value out, executives and board members will lose their jobs, and/or lives."

"Hm. Okay. Hopefully that all goes alright. So why did Ennis call you?"

"He needs another nuclear warhead."

"To . . . ?"

"Blow up the moon. Some of it. Sylber Giles is operating out of a patch of desert there."

"Why in God's name would you trust him of all people? It sounds pretty ridiculous, even for someone like—"

"I've seen the orbital vessel they're building to occupy Tyche. I think Ombudsman and McPherson have probably already jumped on it and whatever is in its orbit to mine the shit out of that whole system in the few years they have before it's flung back out into the galaxy."

"This is the ship you went looking for when you totally betrayed me and nearly got me fired?"

"Right."

"So when they came and asked me about those directories . . . "

Otis finishes her thought, "They were interested because it related to Behemoth, and in turn Tyche. Which means whoever grilled you is a part of the conspiracy, and most likely hasn't given up on waiting for me to reappear in your life."

"What makes you think so?"

"They're following us. Three towncars further back, and a drone up there."

She leans into her window, confirming the drone, sinking feeling in the pit of her gut as realization strikes. "That's great."

"Call Ennis."

A display appears on the dash, 'Calling . . . '

Up ahead, cars stop at a gridlocked ten lane-wide Enforcement roadblock.

"This is it."

"Otis?" says Ennis.

"Imogen and I are in Lauchlap Michael's Res-twelve district in Ontario, about to be arrested by Enforcement."

"Oh no."

"I'll let you know if we get clear, but either way we need extraction. Maybe armed."

"Yeah I'll probably just call Eli and come pick you up then. We'll go get the bomb from there, yeah? Uh, the thing. You told me not to say bomb on a call. I remember that now."

The display vanishes.

"Who's Eli?" Imogen asks as they roll to a stop.

Otis pops his door. "Let's go."

She follows him out of the car, weaving between stopped vehicles with their heads low.

They reach the clearway barrier without being spotted by Enforcement, but Otis sees the drone readjust, then hover there above them. Only a matter of time.

He makes the twelve foot drop without effort, but Imogen lands awkward, rolls an ankle.

Feeling for swelling, Otis looks up at her with a reassuring smile. "Babe we'll meet up with Ennis and I'll take you to Montana. I have a safe house there, surrounded by districts unfriendly to Lauchlap."

"What about you?"

Again the smile, beginning to feel forced. "I've just gotta go do this thing, then I'll be back. We only have a few years before Tyche though. How do you feel about Elysium?"

She shrugs a little, saying "Hm," then tilts her head, softening to the idea. "It's probably good for a kid. At least the first one."

Otis stands, giving her a brief kiss as he tugs on her hand. "Let's motor."

They use the clearway as cover, passing a small homeless colony while they follow its path deeper into Lauchlap Michael land, toward a junction of concrete pillars and tangled overpasses.

Imogen limps at a jog while Otis runs ahead slightly, clearing the way, ready for any motherfucker to come try it on.

handover

.Orson Unpartitioned.OM/UP3578—Luna.2169.06.29.18:55

In Lunar desert three hundred clicks east of Orson Bay, Ennis, Otis, Keo and Eli stand around an oversized alloy suitcase.

"It's too big," says Ennis quietly, "we're about a hundred over what we wanted."

Eli cuts in, "There's nothing around but desert. They'll evacuate everyone the moment they receive the threat."

Ennis turns away, scanning the horizon.

"Ennis . . . this is going to work. My people are ready to go."

Keo packs chains back into his truck. "All yours, guys." He has a quiet word with Otis then drives away.

Ennis returns his gaze to Eli. "You seem willing to risk a lot for your cause, Eli." Ignoring the bait, Eli wrestles with the weight of the bomb, lowering it onto a sled. "I'm wondering what that is."

Before hauling it away Eli says, "Call Sylber at ten past five. Tell him he has four hours."

"How will I let you know if he sends us the location?"

"It won't matter to me either way. This thing goes off at nine, tell your friends to avoid the moon today."

carry on

.Elysium Residential.GP/EH982—Mars.2169.2169.06.30.00:10

At first Ennis finds no one home, so he sets down on Weaver's couch, watching Antarctica, eating mixed nuts from the table.

A low, gutteral sobbing seems to come through the ceiling.

He needs to buy weed while he's on Mars.

As the sobs grow louder Ennis realizes that it's coming from Weaver's bedroom.

He looks at the darkened room, then the front door. A long time deciding.

The frosted glass door slides open when he nears, revealing Weaver on the bed, his back to Ennis.

Weaver's ribcage rises and falls rapidly, but silent now.

"You okay buddy?"

Nothing.

"Weaver?"

His head turns slightly, but still Weaver doesn't speak.

Standing over his best friend in the dark, Ennis says, "Weaver I'm sorry I killed them. I'm so sorry for that. I'm just . . . really . . . sorry now."

He returns to the living room, leaving a jumbo pack of Morpheze syrettes on the coffee table before trudging out.

burn town

.Quarantine Gate WA90—near-Earth orbit.2169.06.30.06:19

From Quarantine, Ennis pilots his rented Visar into atmo before letting reentry systems take over, giving European coordinates to the nav.

'Restricted', comes its response.

He grabs the stick again, taking it in manually.

Nearing the edge of Rotterdam airspace at full burn, descending rapidly, nose afire, an alert appears on primary HUD.

Ten clicks away a pair of automated sentry rigs swing their AA cannons south, tracking the rental's approach.

They launch a brief gauss volley but cut off almost immediately as their target descends below skyway markers.

In the cockpit, Ennis instinctively ducks when rounds detonate above.

Flying low to the ground, rattling traffic as he weaves over and between, nothing is left standing between Ennis Pryce and home.

Amsterdam, 2169

Geiger alerts vibrate on the PDA in his jeans when Ennis steps out onto Amsterdam soil.

It took twenty-one years, but he made it. In all those meetings, the mediation and pointless appeals . . . he was dragged from the room screaming that he would be back. He would find a way, he told those unmoved stares.

But now it's not the same. Amsterdam has lost a lot of its earlier charm.

Ombudsman never bothered to clean it up. Just sealed it off from the world. Most of Rotterdam was abandoned also as fallout dispersed downwind, the whole outer ring soon resettled by unallied masses who had bred and somehow lived here between walls. Normal lives, but with the footnote that their whole world is a genetic timebomb.

Behind him, the skeletal remnants of a concrete bridge reaches tentatively out over the moat ditch, its crumbling support structure half-collapsed into radioactive pools.

Ahead, the city center, its transport hub, and of course ground zero.

He shuffles down the crater's gradual incline, aware of human bones at his peripheral, buried in black glass.

No one even came here to bury the dead.

Ennis says, "Call Apple."

The PDA connects them on audio.

"Morning," she sounds distant, distorted. Radiation all but wipes his signal.

"Leverage is completely useless to us anymore. With Mailer I mean. All we really need is for people to know, so they can plan."

"I can barely hear you Ennis?"

"Get it out to enough people and it'll reach the outer moons."

"Ennis? Hello?"

"I'm going to go and find Grace. I'm going to bury her, and her family. Then we'll hook up with Otis, and tell everyone that 'lyche is coming."

"What? Where are you?"

industry way

.Gaul Commercial.MG/CBD01.2169.06.29.21:01

Keo Virak leaves his ride in an absent VP's spot right out front, buttoning his jacket as he hurries to the lobby of McPherson's prized Gaul Tower headquarters.

Riding the elevator to 990, he tells his PDA to call Goseph Mailer.

"Keo. This is odd."

"I know it's been a while, but I have something new for you."

Silence on Mailer's end.

"Don't try tracing this call," says Keo, "I'm on a boat."

"What have you got for us?"

"I know you spoke to Otis about tracking me down, and I'm pretty sure that you're keeping him close because you've figured out he's a threat to Ombudsman. You just don't see how yet. But I'm up to speed on what's been going on. The whole time I was off your radar, I was tracking Otis."

"Why?"

"Because him being a bigger fish than me is my one option out. I know how fucked I am with you guys. Tyche is about the only place I could go, and even then, I'm sure you'll have that one all stitched up in a matter of years."

"What's Otis been doing?"

"Ever hear of Ennis Pryce? How about Apple Zenith?"

"Maybe."

"These three have constructed an ad hoc insurgency and plan to come after you through Sylber Giles. They know about Tampa, and I think they're going to tell people by hijacking a feed network."

"Hm. From what I hear, you recently supplied another warhead to Pryce."

"Damn straight I did, earlier today."

"Why?"

"I don't have a dog in this fight, Goseph. I'm just trading as normal. Trying to build up a little leaving Earth fund."

"This could hurt your chances on a deal."

"No, it really won't. I'm going to deliver all three of them. You need me this time. Me personally."

"We can probably reach an arrangement."

"Hells yeah we can. But I want more than kickbacks and immunity."

"What, exactly?"

"Guess."

"Not a moon."

"Moon!"

"We'll need to get back to you."

"I have all the time in the world."

chicken

.Glen Canyon.OM/RES21J—Luna.2169.06.29.20:11

Sylber Giles watches the clock, waiting for bunker four to autoseal.

Jesus. How are they going to let this happen? There are more than just antiquities down there.

Part of him wants to pick up an assault rifle and roam the halls until he comes across Ennis Pryce, and the rest of him knows how close to ridiculous that scenario would be, given what a pussy he is.

He's a born executive. Fighting is for those with an affinity for hands-on violence. His brand of violence may at first appear more restrained, but is mostly brutal, even fatal.

He doubts Ennis can even get in here. Unless he knows about the caves.

This isn't just a massive coincidence. He spends a total of three days here every year.

Why would Goseph Mailer ask him to run a recovery drill this weekend, if not to set him up?

Are Ombudsman finished with him already?

Running another pass on the motion sensors, one of the thumbnailed displays maximizes, showing Eli Lylo entering the facility through its adjoining cave system.

Oh fuck, is his first reaction, as he sees Eli dragging a sled that could only be one thing. But he reconsiders almost instantly. Eli is a soldier. A mental midget, next to guys at Sylber's level.

Eleven seconds remaining on the countdown, Sylber exits the bunker, a pang of anxiety as he watches the door seal forever behind him.

crazy bomber 2: operation moonfuck

.Departitioned Uninhab.OM/RES404.2169.06.30.11:03

From his vantage point atop the ruins of Gunderson Interworld's former headquarters, Ennis has spotted an irregularity in the valley that lies between the lone tower and Amsterdam's northern border.

Here in the city's crater, most of its underground transport infrastructure laid bare at his feet, Ennis has climbed one of the few remaining wireframes for a better look at the hypocenter, trying to identify the bomb's position when it received his signal.

But now the valley has caught his eye. A pocket of tall trees—unreasonably tall even, reaching skyward from the old growth forest that lines the canyon wall.

It's enough to move him on, the issue of ground zero all but forgotten now that he sees what a mess it is. There are no answers here, just radiation.

Downclimbing, careful to maintain his points of contact on charred, grimey steel, Ennis decides to investigate the trees, hoping that he will at least find some food there.

Coming out of the crater on its north side there's the sense of cleaner air, buildings with a wall or two remaining, only the occasional skeleton.

When it picks up signal again, the PDA immediately rings.

"Yo."

"I'm standing here with Sylber Giles, he wants to talk to you."

"Okay put him on."

"Ennis?"

"Yeah."

"Whatever they told you was down here, they're full of shit."

"They told me it was a QAARP array. They said you caused an earthquake with it."

"We triggered an earthquake yeah, but not with an array. There were depth charges in the magma flow under Tampa Valley. Antimatter. Take a look at the PDA you cracked, Ombudsman had to make arrangements to physically extract my guys on the day."

"Yeah I saw that. I wondered."

"A team of twenty-four wiped that place clean in under an hour. It came straight from Goseph. How else would we have access to the hardware?"

"So what are you hiding up there?"

Sylber pauses, refocuses. "There are artifacts in the caves. They're older than humanity on Earth."

"Okay."

"Stone tablets, paintings . . . astronomy. It's how we knew where to look for Tyche—they called it Marduk. They plotted its orbit through twenty-two passes of Sol."

"Wow."

"From the drawings though, they appear human."

"Weird."

"Tell Eli to disarm the bomb. I'll give you everything they found here."

"All I want is the exact trajectory of Tyche," says Ennis, "give Eli the crypkey and I will definitely tell him to let you go."

He looks at his PDA a moment, q-mail alerts appearing in quick succession. As he goes through the data, he can hear the PDA being passed back.

"Eli?"

"Yeah?"

"We're good."

recall

.Glen Canyon.OM/RES21D—Luna.2169.06.29.23:08

Riding a ULH-44 Jaguar's skids as it crosses the Three Rivers ten feet off the deck, Peter Dixon feels the pat on his shoulder that tells him it's time to go.

The helo slows just enough that it's messy, not fatal when he leaps.

Derek Walsh comes down with a soft thud too, breathing moon dust while he checks his kit.

Other two-man SEAL teams will enter the cave system from its north and east. The objective is straightforward enough—locate the nuke and disable it, or plant a marker and get the fuck out before the orbiting carrier *Ramirez* tears it apart with Kinetic Storm.

The gravity seems lost on Walsh, or he's better at hiding it. Dixon is beyond shit scared. Not because he's soft. Anything but. Like Walsh, Dixon is a combat-hardened Ombudsman SEAL, going back to a time when it wasn't even called Ombudsman.

Unlike Walsh, Dixon has dreamed since the age of fourteen that he would die on his twenty-eighth birthday as a mushroom cloud rises over Earth's moon.

It is the single image that has haunted him, and comes with a certainty unmatched by any other notion.

Even now, feet padding lightly over Lunar desert, he can't quite believe that the day is upon him and that it could have played out exactly as he always told himself that it wouldn't. The benefit to having his nights consumed by terror is fully realized in his readiness to deal with its pending arrival.

revision

Sylber watches Eli for a reaction as they enter the main hall, convinced that the full impact hasn't quite hit him.

"You get it, right? Why this is so significant?"

Excavation equipment sits recently-abandoned in the underground structure, precisely cut marbled walls lit up by huge floodlights, focused on twelve carvings that to Eli look like stations of the cross.

Pointing at one, Sylber says, "This shit is epic Eli. The creation myth talks about a water world that orbited between Venus and Mars."

Eli begins setting up scanners, laid out in a small triangle at the hall's dead center.

"Marduk was a brown dwarf drifting across the galactic plane. It captured nine planets and their moons in a wide binary orbit with Sol."

"It says all that on the walls?"

"It sort of does. A prehistoric interpretation of science that we went and spent the last fifteen years verifying."

"Go on then. What else does it say?"

"One of its rock giants ripped the water world in half as it swept back through our system about four and a half billion years ago. The crippled planet captured a moon though."

Eli checks incoming data from the scanners as their sensors crawl every inch of the enormous cavern.

"In the billions of years that followed that moon cleared the debris field, and stabilized the water world's tides and orbit."

"The big twist," says Eli, "is that we're standing on the captured moon."

Sylber nods, hesitant. "Luna, yeah. And obviously the other one is Earth. This moon belongs to another star system, and they looked like us. The drawings look like people."

"Well that was what I would have guessed." Eli walks around the ancient site, over to the carving of planets colliding.

"Eli, do you have any idea how huge this is? Mailer is probably setting you up for life, but what's the play here? You're going to destroy the first tangible evidence that mankind is not only not alone, but not unique to this system?"

Eli sits Sylber down with his gaze. "In most of the mystery schools and ritual houses, these traditions they practiced would rely on poorly interpreted ancient wisdom, facsimiled and recycled to the point of philosophical collapse."

One of the scanners chimes, then the others, all three strobing. Eli turns back to Sylber almost sad.

"But one of them contains meaning to this day. One retains the vital clues, and the untainted message. Which isn't to say it's the one path. Just a correct interpretation of the law that governs all paths. Handed down from somewhere beyond any source the others would ever tap."

Sylber can't help a sarcastic grin. "Where are we going with this Eli?"

"Now I have a holograph, we don't need the chamber."

"No! Eli, no. We need it because we don't know how they made it. Or what it does. Our physicist said it's laid out like a huge weird tuning fork. That's got to be something crazy right? Maybe it's a weapon. Tell Mailer that. Tell him it could be a huge fucking energy beam that shoots out of the moon and we wouldn't even know because we haven't scratched the surface. There are levels beneath this one, we can't get to them through the stone blocks."

"Well," says Eli, his voice echoing in the chamber as he gestures to the bomb, "I have just the thing for that."

"Eli, tell him. See what he says okay? Call Goseph and say Sylber thinks it could actually be some sort of weapon."

"It's not." Eli repacks the scanners, shouldering the bag. Sylber is quiet as he watches, brow creasing, finally getting it.

"You know exactly what this place is."

"Yes."

"Why are you here Eli? Are you even working for Goseph?"

"To scan it. To make sure no one else does."

"You're still going to set off the nuke?"

"I am. But you don't need to think that far ahead." Eli has produced a pistol, the sights catching Sylber's eye as they blink on.

"Whoa! You're not gonna tie me to the bomb or something?"

"Why?"

"So it blows me up with my moon base. Like a movie. It's poetic, you know?"

Eli looks around, processing. "Yeah okay. I don't see why not."

"Well I could escape."

"Look dude do you want to get blown up or shot in the forehead because I could give a fuck really."

"Blown up I guess."

Nine

momentum

.Morocco Worship.MG/RC96—Africa.2169.06.30.07:14

Otis and Keo meet in the square amongst worship districts in McPherson's outer spread, favellas lining the valley floor beyond churches and mosques that take pride of place on the mountainside.

Otis tosses a coin into the wishing fountain, returning his attention to the view. "So . . . what?"

They're both distracted as shouting and chanting erupt from a church built into the valley wall, a twenty-strong group of Christians carrying their life sized crucifix up the narrow stairway, past Otis and Keo, continuing loudly to the top of the mountain.

He must remember, thinks Otis, to buy Imogen a gift.

When the commotion has all but subsided Keo looks Otis in the eye and says sharply, "You probably already know that after I left you on Luna, I called Mailer."

"I didn't know that, but I appreciate your candor."

"He has intel linking all of us to the nuke."

"Obviously." Otis catches the bulk of a weapon in Keo's left armpit when he shifts his grip on the rail. "I heard you also visited McPherson HQ."

It catches Keo off guard, not long, but enough that the concoction of a plausible story is evident in averted eyes.

Looking back at Otis, "I've been booking time on the feed networks. McPherson own two thirds of media."

"You've been listening to my calls then."

"Yeah. Apple wants to go public about Ombudsman's part in Tampa, you're not so sure."

Otis doesn't look at Keo, just senses the body language at the edge of his peripheral. Keo tenses slowly, becomes aware of it, consciously relaxes.

He asks Keo, "What do we need to do?"

"I've arranged a broadcast point on Mars, right on feednet's backbone, a thousand clicks from any city."

Otis can't help a smile, almost giving the whole game away before rolling it into a gesture of hubris, taking on an air of misplaced confidence. "Sounds perfect man."

"It is. You'll have at least six minutes before Enforcement can physically intercept you. They route feed to every affiliate from this one relay, and the signal can't be jammed remotely."

"Send me the details, let me run it by Apple and Ennis, if she can find him. I need at least a few days." Otis begins walking back to his cruiser.

"You can't look at it today?"

"Gotta run some errands."

Keo snorts. "Yeah okay. Say hi to your girlfriend."

"Fiancé," shouts Otis over his shoulder, boarding the craft. "You've been hacking my calls, you should know that."

decider

"Yeah Keo is full of shit," Otis tells Apple on an encrypted call, minutes after crossing the coastline. "It's the perfect ambush. I'd say he's promised us to Mailer already, which means we can fuck him double if the opportunity presents itself."

"Hey Otis can you do me a favor?"

"Sure."

"You're on Earth right?"

"Yeah."

"Can you go to Amsterdam and bring Ennis to Mars?"

"He went to Amsterdam?"

"It seems like he's on foot, headed toward the north wall. There's a lot of radiation so you'd need a suit."

"Is he a suicide risk or something?"

"I don't think so. He said he wants to bury his girlfriend."

"She was there?"

"Yeah."

"Her ashes are gonna be blowing around the Pacific by now."

"Don't . . . tell him that, when you go. I'm sure he knows, but don't say it like that okay?"

"Leave it with me," says Otis.

nearly there

.NAPLES RESIDENTIAL.HJF/EH54677—EU.2169.06.30.09:40

Otis climbs the stairs to his third floor failsafe apartment, on hands and knees for the last few, leaning his head out into the hall at ground level, spotting two men lingering by the lift with its call buttons unlit.

Shoot first, check IDs later. He fires a silenced round into the first possibly-innocent-bystander's chest, missing the second when the guy jerks suddenly to catch his friend. The second one goes for a weapon but Otis is already rushing him, squeezing off two more, right in the torso.

Rounding the corner to his right Otis can see a third goon come from the apartment, alerted by the commotion, chewing up hallway with a BA-90 SMG, the silence finally shattered.

Otis backs into the lift bay, its width barely enough to take cover, peppered by debris as the gunman fires poorly-steadied bursts. Otis pops the chest button on his dress shirt and fishes around inside Kevlar armor for grenade sticks.

Biting the cap off, he skips one down the hall, ears popping when it goes off.

Despite the steady hiss, Otis can feel footsteps approaching fast on wooden floorboards. He swings the pistol out and just starts blindly firing, rewarded by a loud collapse, another short burst of BA-90, then silence.

Stepping tentatively back out, weapon raised, Otis stalks into the apartment, clearing it quickly. He finds only turned-out drawers plus his safe, blown open and empty. More will come soon.

Into the kitchen, Otis moves the fridge a step at a time, then kneels in front of the safe door that looks like a dirty wall.

"Voiceprint, Otis Wilmington."

It slides away, revealing a rifle, shotgun, three handguns, and two printed codebooks. An alert appears by his head, flashing 'Proximity[!]'. He unfurls a duffel bag and packs everything carefully, mentally rehearsing the next twenty minutes.

Otis is out the door and up the stairs a moment before the lift doors open. Six men file into the hall, not shy about their weapons.

dog fight

Otis Wilmington is being followed.

They were there with him in Boston, two hours ago. In Shanghai, and Rotterdam.

Now they're shadowing his run into Amsterdam airspace forcing Otis too high to avoid the wall's sentry cannons—legacy hardware powered geothermally, designed to survive the nuclear blast that killed anyone who knew how to shut them down.

The first gauss volley is easily avoided, but Otis knows that his maneuvering has already given it enough data to box him in with its next three attempts.

The only hope now is that his bulky, well-armed pursuers present a more inviting target.

A second cannon swings toward him from the southeast, firing also.

He goes for unpredictable, aiming the Bosworth's nose at it to avoid the barrel by eye, twitching the stick every time it launches magnesium bolts that strobe the interior hard white when they skim by.

The first cannon has picked up one of Keo's goon squad transports, leveraging it into a sharp, evasive roll.

The second gunship is still on him, diving also, missile alerts spooling on the Bosworth's HUD when the THD-08 'Bluebottle' unloads its air-to-air arsenal on him. Otis drops countermeasures as his Bosworth buzzes the cannon, flares landing on the wall and platform around the sentry.

Putting Amsterdam's wall between himself and the missiles, he cuts lift altogether, letting the craft drop, lighting the engines again in time to soften its landing in the interior moat's radioactive mud pools.

Above him, the Bluebottle's missiles slam into the wall, pulverizing the sentry in a series of small detonations.

A growing cloud of falling dust obscures the Bosworth, frustrating the bigger ship's attempt to get another lock. Otis guesses that the pilot will grow impatient as radiation cooks its sensor farms, potentially risk drifting in closer for a better read. He punches the harness release and runs to the rear, pulling down a Stinger case.

More bright orange flashes as the Bluebottle begins launching air-to-ground at the first sentry from somewhere within the wall. He can hear the cannon firing, connecting with some of the missiles, but not enough to save itself.

Punching a dose of Thyrox into his lymph node, Otis splashes out into the mud, hauling the heavy launcher, looking for line of sight. The Bluebottle surprises him though, cresting the wall from its outside, having doubled back impossibly fast.

Its small-arms gimble finds him sooner than Otis can get the Stinger on his shoulder, lead slugs pounding into him at the waist. He staggers backward, focused only on keeping the launcher level, eye to scope. Too close for a lock-on and losing feeling he lines the gunship up dead center then fires, knocked back swiftly against the Bosworth by its immediate detonation, a series of shockwaves slapping at him when the engines and tanks blow.

The Bluebottle drops like a rock, coming down right on the wall, not hard enough to kill its crew. Three to four, he figures, they'll be stunned but agitated, probably with sidearms.

Suddenly aware of himself slipping into microsleep, Otis forces every ounce of will to his numb legs, inching up the Bosworth's hull a shoulder blade at a time, pain radiating everywhere, drowning itself out.

He rolls along the alloy skin back to the side door, holding his shattered right hip, a crimson racing stripe smeared down the Bosworth's skin.

Pieces of the wall splash into nearby pools, silence creeping in when the gunship's turbines wind down. While Otis stumbles into relative safety, one of Keo's men climbs out of the teetering Bluebottle, lugging an assault rifle and ammo box.

The hammer of gunfire on the hull from outside barely registers, Otis can only see the cockpit, working himself fist over fist on the zero-G rail.

"Autopilot," he croaks, the systems coming out of standby. He collapses at the cockpit door, bleeding on the rug there. "Get me above that gunship."

The cabin shudders, engines struggling to light in the mud.

Facedown on the floor, pulling together the sheer focus required for one last set of actions, Otis whispers, "Please."

Keogate

“Keo, it’s Jonathan Maloney.”

“Jon! Hey. Listen, ah . . . what did I tell you just the other day about *ever fucking calling me directly*?!” Keo screams into his PDA behind the stick of a small Warrior class cruiser. “You’re stepping on three and a half years of planning you dolt piece of shit goddamn rookie!”

“It’s over anyway,” grunts Jon. “We lost, Keo.”

A chill that has grown all too familiar since the day Goseph Mailer took an unofficial interest in his affairs. This may be serious.

“Where are you right now?”

“Just inside Amsterdam’s wall.” Jon sounds like he’s struggling to breathe.

“Jon are you hurt?”

“Yeah. We thought we had him, came in low over the sentries and lost visual for just a second.”

“What do you mean had him? Your role was observational. You were to wait on Ombudsman’s go before *fucking moving in!* Remember Mars? Remember the whole ambush scenario? Ennis Pryce and Apple Zenith?”

“Shut the hell up Keo, I’m doing you a favor here. I’m dying okay? Our orders were to hang back, or fire if fired upon.”

“He shot at you? How? He shouldn’t have even seen you if you were doing—”

“Wilmington surprised Mel and Ewan at one of the apartments on your list. He shot them both, then Jason. We heard it all go down on radio, but couldn’t get up there in time.”

“Fuck!” Keo slams the dash, HUD degaussing. “FUCK!”

“We followed him into Asia, holding way back . . . he was onto us by then.

In Shanghai he got lost in traffic and came back around, forced our spotter bird into a cargo ship then parked over Fuller's shareholder hub shooting at us with a scoped rifle from the side door. We couldn't even run TargetOS in their borders, our mounts all locked down. He knew exactly where to bring us."

"Tell me he doesn't know I sent you."

"No. But I think he's outside. Paul went out with a machinegun and hasn't—hang on."

A scratching at the other end, the hatch sliding open, then gunfire, two weapons.

Keo is aware that he shouldn't. But he has to know. The silence growing painful, he can't even help himself.

"Jon?"

"Keo," says Otis Wilmington, a little breathless, pained sounding. "You should have sent brand name soldiers."

crossroads

.Montana Shareholder.LM/SH0010—North America.2169.06.30.17:16

Imogen's home is a lot like he expected. Nice smelling, though stuffy, with dust particles that hover in afternoon sun.

There are no photos of Otis. None of his things here, only one toothbrush in the bathroom.

Sitting on the orange blanket-draped couch, Keo takes a long breath, trying to ignore but at the same time define the heaviness he now feels.

It's a kind of nostalgia, he realizes. This room reminds him of something very specific. It seems to trigger a long forgotten but overly familiar string of subtle emotion, derailed by the suddenly ringing PDA in his hand.

"Hello Goseph."

"I just received a log of your call with Jon Maloney."

"Yeah."

"There's a plan B?"

"Sure."

"Go on then."

Hearing a car pull into the space beneath, Keo gets to his feet, moving back to the bedroom.

"In about forty-five seconds I'll have Imogen Harper. She's the fiancé."

"For leverage? No."

"Why the fuck not?"

"What are you going to do, hold her until the day he's no longer a threat?"

Keo hadn't thought further ahead than putting the gun to her nose and calling her a cunt. "Okay . . . so what? You guys are just gonna retire me now?"

"Where are you?"

"In her house, man. Should I run? She's literally about to walk through the front door. I can make this work, I know how to play Otis."

Pause. "Kill her."

"Huh?"

"To show him that we can."

"Jesus."

"Do it, and we call it all even."

"She's coming." He hears the door open and close, a PDA being laid down on the kitchen counter, groceries going into the fridge.

"Wait," says Mailer, doubt in his voice.

"Yeah?" Quiet, hopeful.

"Make it look like suicide. That will screw with him more than the implied threat ever could."

Hiding in the dark, watching Imogen approach, Keo ends the call with his thumb, raising the Glock .34 on autopilot.

Her life for his. In any other situation, he couldn't shoot a woman in the head. But he knew going in, there would be unpleasant days. It just took a long time in the business for things to get this ugly, and maybe he has softened since then, or at least had hoped it would never come up. Consciously, Keo pulls back. His mind lingers in a comfortable place, far from what goes on here.

shipwrecks

.Departitioned Uninhab.OM/RES404.2169.06.30.12:34

Beneath taller trees, the wreck of a downed cruise ship, laid out on the canyon floor, half-decimated, charcoal black on its near side.

He went on one of these cruises once. They would ferry domestic tourists about the border, pointing out neat shit which can be seen from the island and therefore need never be visited in person.

There must be at least a thousand dead inside. Not even one per cent of Amsterdam's official death toll.

Ennis crumbles. Vomiting, crying, pissing and shitting his pants all at once. His only thought, refreshingly absent of all other thought, is that this would be a fitting end to his story.

Laying on the forest floor, Ennis keeps waiting to pass out but can't.

no news

.Lisbon Shareholder.AJ/YY003.2169.06.30.10:10

After she watches them pack the last of her boxes into the moving truck, they have Apple swipe a waiver then it drives away headed for the tunnel. Her limo pulls in a few minutes after, the chauffer taking all her luggage with a brisk nod.

Crawling through airport-bound traffic Apple checks again for q-mail or a missed call.

This is probably the last time that she will be on Earth.

She tries to soak that up but it means nothing. Looks around, imagines all of it destroyed, but almost feels glad? In a rare moment of devil's advocacy she wonders if the best thing for humanity at large isn't to destroy the seed of corruption.

But of course it's too late. The seed was planted on every other world long before she even existed. And who the hell should get to decide that? She wonders, given the means, motive, and right set of circumstances . . . would it really be so hard to destroy a city, a planet, a person for something you believe in.

Page one

Flying over Amsterdam's north canyon at a crawl, Otis can see a lone figure swimming in the river below, losing sight when he crosses the tree line. He tightens the approach to land by the decimated wreck of an old Collins class Ferry.

Taking shallow, pained breaths, Otis places the systems on idle and unbuckles, beginning to feel his midsection shoot out pain in a way that tells him something isn't right downstairs. He watches Ennis step naked from the water, drowned cat-looking with the long hair and beard.

Ennis watches the cockpit for a curious moment, then meanders over, offering Otis a wide-armed shrug while he goes for the side door, pausing when he sees the bloody hull.

Leaning inside, "Otis?"

"Yeah." Little more than a grunt.

"Is this your blood here?"

"Yep."

"You okay?"

Silence, then, "I might need help."

"Okay . . . that's cool." Ennis looks around the Bosworth's interior, stepping in. "Holy balls, that's a lot of blood."

Otis strains to look back at the door. "Why are you naked?"

"Do you have clothes I can borrow?"

"What happened to yours?"

"I shit myself. I had to take them off."

"Ah . . . yeah okay. There's a closet right in back by the lav."

Ennis enters, going back there.

"Hey Ennis, don't shit my pants too okay?"

"I think I'm alright now. Probably just the radiation."

"You should take some Thyrox."

"Nah it's all good."

Otis feels his arm grow heavy while plotting the course to Elysium.

"Do you know how to dress a wound?"

"Sure." Ennis walks back through the interior in suit pants and t-shirt, still barefoot.

"I took a slug or four to the hip, I'm wearing a vest but it feels like shrapnel got in."

"Show me," says Ennis, helping him up.

Otis yelps, the pain becoming apparent now that he's had time to settle, and the shock worn off. With tears in his eyes he says, "Jesus man."

Ennis supports most of his weight as they hobble back to the cots, sitting him down.

Otis peels off the wet shirt and pulls away the tight body armor for the first time, leaning back to look at the damage. His chest is a patchwork of speckled blue and red, growing darker and deeper down to the destroyed right hip.

"Fucking hell," says Otis, feeling light-headed again, "looks like a sack of broken glass."

"There's no shrapnel here. You're feeling bone fragments sticking out of your hip bro."

Ennis begins cleaning the wound, best he can with limited Gemmadine and gauze. Otis watches briefly then lays back, listening to the soft sounds of triage until sleep sets in.

mercy flight

Ennis maneuvers the Bosworth into a holding pattern beneath hyperspace gate BLI-199 in high Earth orbit, waiting for clearance. In the copilot's seat Otis shakes his head slowly, whispering something inaudible, one of the half-chewed Morpheze in his limp hand.

The cruiser's sparse first aid kit won't be enough. He'll need medical attention before they reach Mars. For a moment Ennis considers turning them around, maybe hit up one of the less Ombudsman-friendly hospitals in Auckland or Fiji. But that would put them on Earth when shit hits the fan.

His ringing PDA breaks the silence, ID flashing 'Apple Zenith'. He surprises himself by silencing it, swiping the alert away. A small part of him is giddy at the thought of Apple worrying about his wellbeing. Another part, maybe larger, is momentarily excited by the idea that he would care what she feels toward him. Her of all people. Imagine that. He hopes this doesn't foreshadow some stupid fantasy of ever melting the ice there. The ridiculousness gives him a feeling of, shame? Not the normal shame.

What could he really offer anyone, by way of friendship, romance, comradery of any kind?

Ennis no longer has the faculties to love a person, to feel things for them and make sure they feel happy. A hug or kiss would be as empty a gesture as all those Vanderbolt specials.

vmail

"I lost track of you in the canyon . . . did Otis find you? My flight was delayed so I'm late to Mars. Just spitballing, but do you think we could trick Mailer into broadcasting it himself? We should all sit down and lay things out a bit more solid, this whole deal is starting to feel kinda fuzzy.

Alright . . . well . . . I'll see you guys there."

safety card

.Lisbon Transit.OM/ISOL05—EU.2169.06.30.15:24

Gently lifting off the tarmac, the big passenger ship lining up its exit corridor, morning sun catches the starboard cabin windows, waking Apple from a light doze, drawing her eye to the Boston skyline.

The stewards run through a final cabin check before taking their own seats. Most of the first class passengers sleep.

Lights dim, then the whole cabin vibrates at a soft hum, and they're moving, incredibly fast. The ship's EM field negates any G-force, but the inner ear can tell. She can always tell.

With a final look at North America disappearing into ocean, Apple dons earbuds and reclines, mentally planning the layout and décor of her home on Mars.

When that gets old she plays out a few scenarios that somehow lead to forcing Goseph Mailer into spilling the beans on Tyche, the leaps of logic growing more tenuous with each.

But then there's one. Not fully formed. Not exactly probable. But plausible.

Her mind drifts back to Ennis, wondering if he's been retrieved by Otis.

She tries Otis first but it rings out.

Then Ennis, "This is the vmail of Ennis Pryce. . . . This is the vmail of Ennis Pryce. End. This is the vmail of Ennis Pryce. End message. End outgoing message. Stop recording please."

the direct approach

.Isidis Operations.OM/MARS01—Mars.2169.06.30.19:59

Apple walks into Goseph Mailer's office and sits down, waiting for him to return.

"You don't think that you're surprising me here or anything, right?" She turns to see him enter with a PA trailing. "I had you cleared when your cab pulled up. There's a rumor that we're holding your *luggage* at the front desk?"

"I just stepped off a flight."

Goseph takes his seat behind the desk while the assistant hovers, staring down Apple.

"Well . . . let's have it. Should I be worried?"

"Not at all," says Apple, "Just an ordinary business conversation." Then to the PA, "What the fuck is your problem sweetheart?"

Goseph gives the girl a look that sends her outside, the door sliding closed again.

"You know Sylber gave us the location, right? We have drones out there now, sending back bioindicators on the surface of three moons."

He feigns disinterest, showing her his palms. "And?"

"There's imaging too. We can tell people where to look. When they see it for themselves, it's all over."

"Moot point. You don't have the means to feedcast."

"Goseph . . . come on. It's a matter of months, weeks if we really push. Didn't some anticorps just commandeer a feed network on Ganymede? I know you're Ombudsman and everything, but you're already on the back foot. You're defending a compromised position, and from what I can see, spending all your time trying to fight fires with the alliances that you started. You can't contain this any longer."

"What is it you think I can do for you? What are your terms?"

"We want you to make a choice. Let them hear it from you today, or us tomorrow."

"You must already know that any disclosure coming from us would leave out vital details about how long Ombudsman and the alliances have been aware of Marduk."

"Okay, good. So you've already put this in the maybe pile. You're smarter than I would have guessed." A pause. "I can live with any spin you want to put on it."

Cautious, "Why?"

"Have you been on Earth lately? People are just going about their lives. They're all going to die because you wanted the big three to have their pick of land on Mars and Triton?"

Goseph masks an involuntary grin, but it's all over his eyes. "There's more to it than geopolitics, this time around."

"Dude I *get* why you did it. I wouldn't expect any less, it's just how people are . . . but it's time now. How much longer would you have waited if we didn't find out?"

He glances away a moment.

"Never," says Apple, feeling her will draining. "Of course. Because all the people who didn't find out through the right channels weren't worth a lot to you anyway."

"It's already too late, and far too expensive, to mount a rescue effort for the majority. We have a strategy in place to remove around fourteen thousand VIPs when the time comes. We ran the projections, twenty-thousand plus and it devolves to mayhem."

"People will work together. The companies will evacuate their own."

"Shareholders, and top tier execs."

"They'll take the employees too. They need experienced people to run the new infrastructure."

"You don't seem to understand. We can always get more people. What's harder to find are the wealthy and influential, with views aligned to ours."

Apple stands. "You'll think about this?"

"If we tell them, there needs to be some guarantee that the official story remains the only version out there. I never want to hear another word about Tampa Valley."

"I can guarantee that."

"And Ennis Pryce," a knowing sneer, "you can keep a lid on that hairy little shit?"

"Of course."

"Then leave it with me. If you see it all over the feed tonight, consider it a deal."

grapevine

"I'm looking for Mr. Wilmington," says the stranger, "it's Rema from Saint Oliver's in Montana, I've tried his personal contacts but no luck."

"Otis hasn't arrived in the office yet I'm afraid. I don't know when to expect him."

"Hm, this is a bit of a gray area, I need to notify Mr. Wilmington of a death, we can usually do so through family but I don't know if it's okay for a colleague . . . okay my supervisor is nodding, are you able to pass on the details?"

"I guess." Then, of course, "Who died?"

the dying room

.Van Casper Emergency Care.MG/DH004—Luna.2169.06.30.21:35

Ennis sits up in the stiff plastic chair bolted to a row of identical seating, straining to hear Kirk Vanderbolt speak above the hustle and wailing of other patients.

Otis stretches out on the floor at his feet, waiting for medical attention, orderlies and nurses making a big show of having to step over him when they rush by. Fuck you, go fuck yourself, Ennis tells each of them with his glare.

Without opening his eyes Otis says quietly, "You should ask them to run your vitals too."

"I'm okay. I didn't feel like there was much radiation back there. I'm very in tune with my body like that. My PDA has been going on and off for a while though."

"Mine too. That's radiation damage."

"Tell you what, if it turns out I'm radioactive down the line I owe you a Coke."

"Ennis why did you trigger the bomb?"

"What do you mean?"

"Why did you do it? You said the whole idea was just to go home."

Ennis watches Vanderbolt's lips move, noticing the dried-out skin, the age lines.

"They stopped me at the border. I wasn't going back that day or any."

"You could have just given up right there. You didn't have to do it. Was it pride?"

No response. Through the veil of delirium, he can't tell if Ennis is thinking, or being willful.

"I read about Luna Phranc, the border guard. Did you know she looked right at the blast? I think they said she still lives near Rotterdam, can't have kids and stuff . . . "

"Because of the bomb?"

"I dunno man . . . I would say that's probably the reason."

Over Vanderbolt's left shoulder, a panel reads, 'Breaking News!'

Standing, Ennis says, "If they caught me, they would have executed me." He walks over to the feed, raising its volume by hand.

"It was just supposed to scare everyone enough that they would let me walk right by . . . she didn't seem to know. She wouldn't get it. No one bothered to tell her I guess. Because if they did . . . jeez. Do you really risk everybody's life just to stop one shitheel living in your fucking country? No one does that. If they'd just told her."

Ennis is quiet again. Then, "Up until she shot me, there was no doubt in my mind that the threat would be enough."

"So what happened?"

"I panicked."

"What? Why even have a live detonator if you never wanted to set it off?"

"They might have been able to tell. I didn't know enough about signals and technology to wire up a fake one. Like the bomb. I read that even if you use the same kind of materials, satellites would have scanned it and seen that it's not fissile."

Otis opens his eyes, looking over at Ennis. "So you're saying now that Amsterdam was . . . an accident?"

Ennis shrugs. "The result was. I convinced myself early on that I would use it if necessary, but over time I realized that would mean killing a lot of people. It wasn't a realistic option."

"That wasn't what you said on Vanderbolt, or to Ombudsman?"

"Because of Mailer."

"He coached you."

"Not directly. I've never spoken to the guy. But he had an alliance lackey waiting when I woke up . . . a statement prepared. They made me out to be a revolutionary. He said it was the only way they could call it a 'necessary reaction to threats of violence under circumstances of war'. Some shit like that."

"Never knew that. Keo did tell me they wanted Amsterdam gone a good while before you came along."

"Apparently—" Ennis suddenly snaps his attention to the display as Vanderbolt's balding head is replaced by deep space gravity maps. "Hang on. What the shit? That's Tyche."

A silence falls over the emergency room as other patients and staff begin noticing the various displays, different networks broadcasting the same images.

Ennis volumes up again.

"—literally going to either crash *into* or come near to the Earth. A world ender, is ah . . . what they're saying. A world *ender*. Is that right? Wow."

Vanderbolt's commentary, over course and damage projections. "We're about to cross live to Isidis for an official statement from Ombudsman's Joshua Lopez."

full disclosure

.Isidis Operations.OM/MARS01—Mars.2169.06.30.21:44

Josh Lopez steps up to the podium and begins reading from his autocue, determined to get through the entire speech before looking at their faces.

A hush falls on the Allied Press gallery, Josh's breath conspicuous in the mic.

"People . . . today I'm tasked with delivering just about the worst news I can imagine."

He shuffles, settling in.

"Three hours ago Ombudsman with the help of key alliance members was able to confirm the existence of a massive dark star in binary orbit with our own sun, a body that we now know has passed harmlessly through our system many times, but once before with utter devastation."

He's aware of someone at the back of the room whispering.

Shut up. Shut the fuck up. You're fucking ruining it.

"As you may have seen in data released by Ombudsman moments ago, the brown dwarf system called Tyche is inbound on completion of a thirty-six hundred year orbit around Sol. Our best minds are predicting a seventy-four per cent likelihood that Earth will experience a direct impact with one of Tyche's planets or moons during this transition."

Josh pauses, still looking down at his hands.

"We have three years to move everybody off world. At two years, Tyche's gravity will begin causing tidal and geological effects on Earth and Venus, resulting in catastrophic quakes, supervolcanic activity, and most likely a pole shift. In four years, the Tyche system will arrive at our doorstep. We don't know what happens after that, as there are too many variables."

Gasps. Now he looks at them. Exactly what he would have expected. He soaks it up, pushing away the image of how his son must feel at this moment, some regret that young Josh II had to hear it with the rest of the world.

But then he's buoyed by the idea that this could be the catalyst that forces his bitch of an ex-wife to really reflect on what a bitch she's been. Especially if she can't get out of debt in time to vacate. That would be a real shame, Claire.

The looks range from devastation to disbelief. For some reason, Josh loves being the guy that gives people bad news, like if they got a ticket, or some celebrity died. But this . . . wow. Better than he thought it would be. And fuck it, he has a house on Mars. Two if you count Marissa's condo. He already has the flights booked.

Everyone at Ombudsman wanted to palm this off on Josh, but it isn't such a shitty assignment. This is epic. This is his telling people the world is ending moment. He will be remembered for the rest of time as an instrumental cog in the machinery that heralded the end of Earth.

Just look at their faces. They all want Josh to tell them what to do next.

"Well, goodnight."

untitled

.Odessa Commercial.MG/HO09—EU.2173.02.17.19:16

Deborah Barclay leaves work at quarter past seven with a good ten minutes to spare before the express to Izmail arrives.

The platform is quiet. A man watches her from the other end of it briefly, then returns to his open briefcase. On the train it's a similar story. Have people just stopped going to work? Should she? The feeds are saying no, continue as normal. All employees will be evacuated in time, according to Ombudsman and her alliance rep, but where are they? The only employee-level evacuation she's aware of are the people who dumped their retirement funds into transport for themselves and their families and all their worldly possessions, then a further mortgage on any spare patch of desert on Luna, or if they're lucky, Mars.

The shareholders are long gone, taken away to new, fertile lands in the days following the initial announcement.

She could probably stay with her foster sister in Elysium, but transport is an issue. Everything is booked solid according to the feeds. A lot of solarlines have wound down Earth services, leaving few options to accommodate a good seventeen billion people around the planet still desperate to escape.

The company isn't going to do jack shit she decides all at once. She can't fly, or afford a pilot, so hiring a ship is out. Her best bet would be to look for a group of people who have already chartered a flight, and get in on that.

the brush off

.Sarajevo Agriculture.OM/DF27—EU. 2173.03.12.16:22

On a low-lying grassed plateau that until recently nurtured dairy and beef cattle, Keo is carefully stepping through barbed wire fencing at the field's perimeter while Mailer's private transport touches down nearby, his crack team of praetorian bodyguards forming a circle around the densely armored gunship.

Two of them move apart to allow Mailer through, falling into step with him on either side. When Keo is within earshot Mailer says, "The board voted."

"And?"

"This is it. All yours."

Keo stops a few feet from Mailer, looking for any sign in the expression of his men that they're steadying for a kill shot. "What?"

"You get Earth."

"Hilarious."

"We need an ambassador down here. Our astrophysicists are predicting a twelve per cent chance the entire Tyche system could pass us by. It's still guaranteed to wreck the magnetosphere but without a collision, Earth is viable again in twenty-five years."

For a moment, Keo is drawn out of the numb aimlessness of the past few weeks by an inner egoic response.

"You want to make me a General?"

"A Brigadier General, sponsored contractor. If planets fail to collide four years from now, we'll need someone coordinating a staggered return of Ombudsman forces to Earth."

"How many under my command?"

Mailer knows now that he already has him. "Seventy thousand. A carrier group. SEALs, Rangers, pilots and infantry . . . the works."

An image of himself in combat fatigues flashes on his mind's eye, suddenly conscious of the extra weight. He'll have to start jogging, maybe lift weights. Then he'll look the part. "So let's say that all happens, you guys come back to Earth before the alliances can reclaim any abandoned territories, reestablish the power base and sell it all back to them . . . then what?"

"You would remain here in charge of world forces."

The deal is done. Keo can't even keep up the pretense of mulling it over. "Yeah," he says slowly, "yeah okay."

"Brilliant. How about a sandwich, have you eaten? My wife packed two for me today."

Keo looks at him for a while. "I can't tell if you're being serious."

"I get that a lot." Mailer walks back to the gunship, tossing back a final, "Don't fuck this up Keo."

Watching it flatten the tall grass then lift off, Keo holds his bloated gut.

Ten

riders on the storm

.Alice Springs Commercial.FE/DK26—Australia.2173.08.09.05:10

When the sun has been below the horizon for three hours they lift off in the Jarhead, its nonessential internals scattered about the edges of an improvised landing field. Jess watches them clear the hills then goes inside to continue setting up one corner of the lobby as a triage point.

The helo coasts dead silent on its EM field, ten meters off the deck, shy of mach one. Umberto pings only once for terrain data then shuts down every broadcasting system approaching seventy clicks out. If he's lucky, their boundary isn't sensitive enough to even register the hit.

The three others prep their weapons in the empty cargo hold, Otis eventually taking the co-pilot's seat, facing him.

"So you're something of a shit hot pilot, yeah?"

"Sure."

"I mean a proper pilot. Licensed to fly something like a rotor craft?"

"Yeah I've flown them."

"Look I didn't mean to give you the impression back there that we were strongarming you."

Umberto nods. "Not at all."

"There are a lot of people being held against their will, by a guy that marched into town and set up camp here when the world started caving in."

"How many exactly?"

"Three hundred maybe. He has them working on an Ombudsman destroyer's reactor core, trying to get it operational so he can flee before the second pass. They're all disposable to this guy, no radiation protection, even food and water are being withheld."

"If it takes three trips to get them all out," asks Umberto, "how do you plan to get them off world before we crash into those planets?"

The man looks out through the forward windows, tightlipped.

"Oh. Okay. So why are we even moving them?"

"That fucker's been killing people down there. Maybe we can get half of them to Luna. Even a third. The rest will at least have their life back, and a chance to survive underground."

They fly a while without speaking before Umberto asks, "Do you know this warlord guy or something? Is he another soldier?"

He snorts a little, turning back to look at his friends. "None of us are soldiers. We're just normal people."

Umberto brings it over the treeline and straight away can see the camp on long range optical, surrounded by concentric rings of razor wire fences and wooden sentry towers.

"Where am I putting down?"

"No, I need you to open the side door and hold a firing position on approach."

Umberto looks across at him. "You're going to fire small arms out the side door? That's the plan? Why don't you just go in silent and pull the people out?"

The guy says, "We need to neutralize the sentries before they can call it in."

"Ten clicks."

"Open the door, slow it right down and bring her in sideways, steady as you can." Otis winces as he gets up, stiffly limping back to the rear, where the other one has set up a long range scoped rifle on the deck.

They lay down on the floor to sight up as the helo's momentum shifts then the side door slides back, blasting the interior with cold wind.

The scruffy guy calls out wind and distance while the cripple fires the first of four subsonic headshots, waking the older man again.

family meeting

.Van Casper Emergency Care.MG/DH004—Luna.2169.07.01.12:27

Apple finds Ennis standing by vending machines, sipping something orange from a plastic cup. The first thing he says, without really meaning it the way it sounds, is that she didn't need to come at all. Otis is going to be fine. He's in surgery now, having his hipbone and a few vital organs seen to.

She tells Ennis that despite the twelve hour flight she still can't think of a way to tell Otis that his fiancé shot herself in the head yesterday.

Ennis sinks back to the wall, lets out something of a laugh.

Sensing the tremble at his collar she steps closer, holds the back of one hand to his cheek. "And what about you?"

Looking back up at her Ennis says, "My PDA broke." Beneath the beard his lips are chapped, pale as his face. Eyes creased, sunken, a different sort of dead than his conventional stoned glaze.

"Have they treated you?"

"For what?"

Apple nods, then goes inside to the front desk.

complications

"Things aren't going as well as we would have hoped," she tells Otis, still powering through the sixth hour of a medically induced coma. "They had to put you under. A few days to heal."

Ennis stands silent at the window.

"Imogen's funeral is tomorrow. Someone should be there in your place, so I'm on the next flight. They tell me there'll be a recording, if you do want to view it somewhere down the line."

She looks over at Ennis then, trying not to give her sentences any weight, but knowing he's already chosen to take it poorly.

"Ennis will stay here with you. He'll tell you what's happened. After that . . . for obvious reasons, I think the three of us should have no further contact. When I leak the rest of Sylber's PDA, Mailer will come back at us. We're all better off disappearing for a few years."

Apple stands, watching Ennis.

"Cool," says Ennis. "Godspeed."

The room is cold. There is a hum to the lights, the ping and grind of machinery breathing, pumping fluids, flushing out Otis.

Leaving, there's a note of impending finality if not resolution.

Technically they achieved their goal. Most likely the three will be hunted across Sol, found and executed within a matter of years because that is the nature of their oppressors. The people who run the universe are simply too large. Too encompassing. This is what we're all left with.

It doesn't occur to Apple until she's in the cab that she has no idea where to go now.

requiem

Within what his team has been calling the King's Chamber due to its similarity in layout to the original, Sylber Giles is handcuffed to railing on the raised walkway surrounding one of the dig sites.

Eli kneels beside him, programming the count, hashing encryption keys, finally punching 'Confirm'. Twenty minutes. Now Sylber knows that he has exactly twenty minutes to live.

"My ride is five minutes out," says Eli, swigging at a canteen before placing it in Sylber's hands.

"Eli, is there anything I can say or do to save my life?"

"Of course not. If there was you would have done it by now. You're that kind of guy Sylber."

Realization is setting in. Indescribable bleakness. The thought of dying here today is enough to make his stomach lurch. He swallows hard several times to avoid throwing up.

Eli exits via the walkway, leaving Sylber alone with the bomb, just out of arm's reach lest he attempt any last minute tampering with the unit itself.

A brief rattle of the cuffs against hefty steel confirms that he won't be going anywhere. No final dash of heroics, despite the nagging feeling that there is some course of action unpursued that would have done the job. Even if he could get free of the chamber though, where would he go now?

Sylber watches it count down, feeling every second grind on the pit of his gut. Complete impotence, a lesson he never really had. Finally, in its three second home stretch, Sylber says aloud, "God, I know I haven't—"

The light is seen first by residents of settlements atop the Glen canyon system.

Luna quarantine picks it out from orbit as a visual spectrum spike, then X and G wavelengths going off the chart.

The mushroom cloud, due to Luna's low gravity and stable atmosphere, hangs there for nearly half a day, witnessed from as far as Deaconshire and Van Casper regions.

benefactor

When they wake him for consent Otis elects not to go for the artificial organs and skeletal hardening, choosing instead to recover at his body's own natural pace.

They tell him it was touch and go. They urge he reconsider, but Otis is only interested in tackling the rigmarole of checkout waivers and the necessary arrangements for physiotherapy.

As he gets through telling the wing administrator that he'll need a referral somewhere close to his fiancé, Ennis can only blurt it out, meaning to wait for a private moment but no longer able to exert that kind of control over himself.

The news doesn't so much shock Otis, as confirm his lifelong suspicion that something truly tragic is perpetually just waiting to occur.

He's quiet of course, long enough that Ennis and the administrator take their discussion out into the hall, arranging for the therapy to be undertaken near Weaver's home on Mars.

reconnoitering

.Restricted—Deep Space.2169.09.06.00:00

The first of four unmanned probes unfurls its science package and optical array at ten million clicks from targeted space, silently sweeping the empty sky ahead for telltale granules of distortion.

Detecting three stellar object candidates, the probe performs a minor course correction, assigning its three companions to the remaining two bodies.

At four million clicks out, the probe has both qdar and visual confirmation of a dark star, hurtling toward Sol.

Settling into orbit around the monster brown dwarf, beginning to pull ground data, the other probes are reporting immense rock giants with moons larger than Earth itself.

Crossing the equator, about to complete a maiden loop before regrouping with its formation, the probe is registering movement on the surface of the failed star called Tyche.

collide-o-scope

At one year from impact the Tyche system's gravitational influence becomes more than apparent on Sol's outer moons.

When Triton's orbit deviates just a few meters the devastating effect on its weather manipulation and terraforming grid results in a media panic that lights a fire under the ass of those who can afford to leave Earth but have been putting it off for whatever reason.

At six months out the only people left on humanity's home world are the miscreants and non-affiliated. Around seven billion of them, just laughing it up, wandering around fighting over the free shit people left behind.

Some of them scheme to steal or otherwise obtain tier two transport off world while others have just come to terms with it. Most, through either a sense of mortal delusion or general naiveté, believe the coming star will pass them by without incident.

And while all of them struggle to make sense of what is about to occur, the truly religious rejoice, for their lord has come to fulfill a pledge long overdue.

phoenix

.2173.08.07.01:37

The quakes ramp up to a hundred minor tremors a day before the big one hits, splitting Earth at her belly, shifting the pacific plate ten miles before the New Madrid fault follows it into frenzy, displacing the great lakes and

Mississippi all in one thrust, a towering inland sea that washes over a few hundred million souls on its way to the Gulf of Mexico.

Volcanic vents open up across the southern Pacific, swallowing New Zealand and the island chains as tsunamis begin to radiate out across the globe, slamming down on the East coast of Australia, scrubbing South-East Asia from its moorings as every tectonic plate on Earth is triggered by the deluge.

From orbit, and safer worlds, they watch. Most weren't even Earthborn, but every pair of eyes would call it home to some degree. Live feed streams out from everywhere that was once a place. Familiar skylines vanish as landmarks give way to nature's fully realized rage.

This is not romantic. There is no joy to be found, even for those who in their fervor anticipated Marduk's coming as some sort of cleansing fire.

Volcanic rock hardens to form fresh mountains and sharp, shiny black islands as the waves slowly lose momentum, Earth settling back into herself, washing the dead into her seas.

Eleven

the fourth year

.Collier Residential.OG/PD112—Mars.2173.08.02.17:18

Weaver is unsteady on his feet. Packing a second hit of salvia into the handheld vaporizer while Ennis and Otis watch, waiting their turn. It occurs to him briefly that he may not have given the first one enough time.

He's midway through pulling the cone when he sees Apple tentatively walk in, pointing at her for lack of a verbal option as the madness takes hold. She goes to speak but is interrupted by Weaver laughing. For a long time Ennis and Otis just watch her while she waits.

Finally Weaver has it out of his system long enough for her to say, "I know I said you'd never see me again."

Weaver gets going some more so she just sits, looking at his strained, red features while he passes the vaporizer to Ennis, who immediately hits it.

Just as Weaver finally sinks into a half-conscious stupor, Ennis begins laughing his ass off also.

"How's it going?" Otis shouts at her. Apple just smiles pleasantly, shrugging.

Ennis hands the vape to Otis, pinching the bridge of his nose, guffawing into the palm of his hand and losing his shit so bad that he can only flounder.

When the room is quiet again Apple tells Otis that she's located her dead sister's missing children. She asks if he and Ennis would be interested in helping her retrieve them from an Odessan refugee camp before Tyche swings around the sun on its outward leg.

bootstrapping

.Collier Manufacturing.REFS/J5—Mars.2173.08.06.23:45

In the idling taxi Otis hands their lone revolver to Ennis and says, "These guys tread lightly to make you feel secure at the buy, because their fourth man is a sniper. Their MO is to arrive about two hours prior, which to answer your earlier question, is why we're here at twelve. It used to be just one guy, and he'd always take the most obvious line of sight. I picked this road because it's essentially a canyon, one main option for their dude to set up. Now, it is possible that in the seven years since I exposed their lack of forethought, they've learned to bring another shooter."

The driver seems suddenly agitated, waiting to be paid, looking around, aware now of the neighborhood's graffiti and level of dilapidation.

Waving the loaded revolver at Otis, Ennis is saying, "Just be careful. Don't get shot."

takeover bid

From the adjacent rooftop Ennis watches Diego's sniper roll out his yoga mat and get comfortable, assembling the rifle from a prone position.

Before he even has the bolt in place, Ennis is straddling the guy at his center of gravity, muzzle to the rear of his skull where nerve endings will most accurately report its shape.

"You can probably already tell what's about to happen . . . we're basically double crossing your boss Diego, none of us having any money as we're all on the run from Ombudsman et cetera. I was probably going to shoot you if Otis had his way but I don't want to kill anyone I don't have to. What do you say I just let you run, straight down the fire escape and to the onramp so as I can track your exit for a good click and a half with your own rifle?"

The sniper nods rapidly.

"That way I know you're not a threat to my friends on the ground. You can leave your PDA here. No shouting, or clever stunts. Yeah?"

Getting off the guy, Ennis covers him with the revolver while they both get to their feet, Diego's man going straight for the ladder.

"Adios." Ennis assumes position on the sniper's surprisingly comfortable mat.

Through the scope, he watches Diego show body armor, rifles, sidearms, and explosives to Otis.

Ennis lines up the crosshair on Diego's open right palm. The way he pictures it is kind of a neat stigmata arrangement clean through the hand. A warning shot across the third and fourth metacarpals. When the scope settles again, shot still ringing in his ears, Ennis can clearly see that Diego is missing an arm.

"Shiiit," under his breath, "that's probably not what we would have liked."

survivor

.Payerla Drydock.OM/NDD06—EU.2173.08.07.06:28

Odessan refugees have begun massing along the border walls where Ombudsman naval yards meet the unpartitioned red zone called Payerla by its former inhabitants.

There are at least a million of them, waiting patiently for now. Probably expecting to at some point be let in, unaware that despite its kilometers of underground tunnels linking drydock repair facilities, food, water and air will become finite resources the moment they seal off.

Oblivious to the border situation, the last of Keo's soldiers are pulling out at Mailer's behest in the territory's one remaining tier three transport.

Everything he was promised is gone now. He will need to fend for himself. He may not be able to talk his way aboard that final ride out of here, given Mailer's well-socialized order that Keo be the man who goes down with his planet, but he would wager his life that there is something in these yards he can get running with enough help.

There are fifty or so remaining mercenaries hired personally by Keo, sheltered with him beneath Odessa for the moment. They're all looking for a way out, each man growing prickly, sitting around waiting for the inevitable call to action.

Good soldiers they may be, none of them will pick up a wrench to kick start the escape.

One of the few he trusts brings Keo to a shelled-out freighter scheduled for dismantling, quarantined in the radiation cage, its exposed core glowing visibly white through the holes in its frame.

As the navy transport lifts off outside, watched by the restless crowd, Keo is finally called to the main entrance over comms.

On loudspeaker, Keo informs the assembled masses that he will need a thousand or so people to work on getting everyone off-world before Tyche makes its return pass. Those who volunteer to work will be given priority safe passage.

voluntary

When the gates open they surge forward ahead of her, as do the people packed in behind Deborah. She's funneled in before they seal it off again, pushed underground at first by the flow of the crowd, then guided with a nudge here and there from soldiers, lining their path through the maze of oversized concrete tunnels.

When they jog out into the open natural caverns holding skeletal spacecraft parts and airframes, abandoned tools and machinery strewn about the place to collect mould, they are immediately herded through to the rear, where an enormous Faraday cage envelops an old Ombudsman naval destroyer with its glowing heart exposed.

When they're all inside the soldiers seal them in with it, one of the men barking orders at those who collected tools on the way through.

A hierarchy begins to form, those with some technical knowhow explaining the process to those with none, before approaching the torn-open engineering section in the ship's aft.

Soon, watched by the soldiers beyond their new prison walls, everyone works to seal off the ship's damaged core, the welders and spotters already beginning to feel the harmful effects of exposure.

Deborah tries to keep in back wherever possible, assigning herself the task of collecting suitable scraps of alloy from the protected side of the crippled destroyer.

stakeout

.QUARANTINE GATE LE02—NEAR-EARTH ORBIT.2173.08.08.21:54

In near Earth orbit, the loaded Bosworth sits dormant behind cover of the abandoned Quarantine station that once serviced Luna gate LE-4066.

The surveillance package welded to its hull takes a snapshot of Odessa's Payerla zone every ten minutes, analyzing troop and resource movements in the wide sprawl of yards. For six hours they've had enough information to go in, but Otis is hesitant, overly cautious since learning of Keo's involvement.

Tension rises again momentarily when a heavy troop transport lifts off from the Ombudsman anchorage and begins to break orbit, but calm returns as it corrects for Mars gates.

Most likely their metrics spotted the Bosworth there idling in the station's shadow, three lifesigns aboard. For caution's sake or sheer lack of interest they moved on. Next time, Otis reasons half aloud, they probably wouldn't be ignored—by a patrol, or an armed interceptor's last minute sweep of the orbital sky. Warnings have gone out all over, claiming that anyone within a hundred thousand clicks of Earth at the second crossing would be rounded up and detained indefinitely. The witnessing of Earth's last moments will occur only via official channels, brought to you commercial free.

Ennis, listening to Ombudsman audio chatter, tilts his head before looking at them. "The last of his troops just bugged out after a warning from Lunar observers that the Pacific is about to overturn. No idea how long, but it's showing signs. Got to be, what, only Keo and the hired goons down there now? Fifty of them . . . fifty-five."

Otis watches live feed of the main gate, saying nothing.

Apple just looks through her window, down at Earth, wondering how she'll go with the weight of her gear and rifle when the running starts.

"Do we even know if those kids are inside the dry dock, or somewhere in the camps? If it's the camps, forget finding them amongst the million refugees before the tsunamis hit."

Apple asks, "Are you scared?"

"Yeah. But it's okay. We've got Weaver."

"Weaver is the man," Otis says. "This thing goes to shit and Weaver will be the guy with a plan, saving the day as per usual. Look at him over there, just sleeping. Like a cat. Lays down anywhere and sleeps. Lulls his enemies into a false sense of wellbeing."

She smiles. "Who are his enemies?"

"Dicks," says Otis.

"Cocks and dicks, yeah," adds Ennis, looking over at Weaver, sleeping on the cruiser's cot in back. "At Octoberfest I was getting arrested, Weaver just jumps out of nowhere and beats the shit out of two peacekeepers."

The closest biway gate's data stream flashes an incoming arrival, busting Otis finally out of his malaise, the next three steps all snapping into focus as one.

"Jughead," says Otis, watching it phase out of hyperspace when the gate's outer donuts shoot it back into third density. "It's Medevac, not navy. Heavy gunship without the guns, but a reliable lifter."

Ennis asks, "Should we follow it then?"

Otis is already plotting a course to match the helo's reentry, shoving them back in the seats. Earth grows larger through the front window while Sol and Tyche rise on its eastern limb.

blocking

.Payerla Drydock.OM/NDD04—EU.2173.08.08.06:02

"Just go in there," says Otis, "start a fire if you have to. Make sure they all leave. Guide them back to the evac and squeeze them into the Jughead right to the ceiling. Its max capacity is, what, one-twenty? You'll get two and change in there, easy. Just find the spaces. It only has to take off."

Weaver returns from the tunnel entrance, a man-sized door sunk into the south wall of the naval yards, inside the main fence but short of its strongest defences.

"We can blow it open," he barks at a whisper, "hand me a tube of composition."

Weaver leaves with it while Ennis swaps his rifle's DU-heavy cartridge out for lighter rounds.

"What about you two?"

Otis stares at the single protruding shape that looms over the whole mess of docks and launchpads. "Keo's probably upstairs in the control tower. I guess I'll go up there and kill him, so we can stop running into him at these things."

Apple stands, checking her straps and holsters. "My contact says Deb was brought in with the most recent group of workers. I'll come down to the radiation cage with you and Weaver."

A muffled bang, then Weaver runs back upstairs to them with his thumb raised, trailing smoke and dust. "I made a hole."

Ennis presses a hand to her shoulder, firm enough that she turns back, annoyed.

"Who told you they're here? I need to know we're not being fucked with by someone with--"

"Eli."

"Lylo? What's his stake in this?"

Apple watches Weaver then Otis enter the blown-open concrete door, disappearing in the shadow beyond.

"He's here too. Somewhere. I don't know why, but he's been amongst the refugees since Tyche passed."

She pulls away then follows through the wall, tasting the bitter ash of charred composition and concrete.

quid pro quo

.Payerla Catacombs.OM/UP0306—EU.2173.08.08.05:49

Eli and Deb scuttle on their heels and asses over smooth bedrock boulders, deeper into the cave, nearly two thousand feet underground now. There is a sense of urgency, Deborah spurred on by Eli's constant reminders that they will drown in the approaching flood if they don't make it through the natural airlock in the next twenty minutes. Fifteen now, shouts Eli over the clacking of smaller rocks matching their pace down the ancient riverbed.

Out of breath, Deb begins to slow, losing her balance when the rock she rests on rolls forward, slamming down on the back of her calf and heel, instantly crushing her lower leg as two boulders slam together.

She can barely scream. It's more of a gasp, a muted moan. Eli stops, knowing what has happened without even needing to look. This will be hard for him.

He goes back to where Deb is already drifting between trying to free herself and passing out from the pain.

"I'm good," she squeaks out, "I'm ok to go, just move the rock for me."

"Deborah."

"I know. Go on then. I probably wouldn't have been able to live with myself anyway . . . we told them we'd go back for all of them."

Eli can't take the time to be delicate, so comes right out with it. "It's not that simple, Deb. I need your marrow. That was the whole reason I pulled you out specifically."

"What do you mean?" Deb can barely look up at him, head heavy.

"There's something happening out on Triton. Your birth mother is involved."

This snaps her out of it, anger flushing pink through her features. "My mom died years ago."

"No she didn't, Deb. Sorry to be so blunt, there were obviously reasons why she couldn't tell you herself, but she survived that crash, and lived on Triton for a long time."

Deb is silent. Eli moves things along quickly, listening always for the roar of water.

"A virus has begun to spread, it started there. Something nasty, on a Solar scale. Monique Zenith has a natural immunity to the pathogen but not robust enough to reproduce in the lab. Their theory is that her proginey would display a more pronounced immunity, something evident in the RNA and marrow."

Deb slowly nods.

"I need your bone."

She offers him her wrists.

"Hang on," says Eli, going around behind her. He rolls the boulder off her crushed leg, slowly at first, then one hard shove with momentum.

This time Deb screams, filling the cavern.

Seeing how bad it is now, he knows that blood loss will take her within minutes. Still, he doesn't have even that long to wait.

"Eli, can you—"

He brings the back of his gloved hand down on her temple, knocking Deborah out cold.

Lighting the small cutting torch, taking her leg off at the calf where the bones are broken, Eli is mentally noting that he still hasn't heard the wave he saw from orbit.

It must have slowed. Which of course means that he has more time than anticipated. He takes an extra minute to cauterize the knee and wraps it with a triage pack, punching a shot of adrenaline into Deb's shoulder.

She blinks awake immediately, heart beating hard, eyes calm as fuck.

"What happened?"

"Can you run?" asks Eli, zipping the leather knapsack that holds her leg.

"I can try."

He leans in and scoops her up, helping her to hobble down the remaining incline.

"If I hear the water coming I'm gonna drop you," Eli tells her, "because it matters more."

"Agreed." Deb feels more capable than at any other time in her life.

rescue party

Ennis tells Apple and Weaver to wait in the tunnel then steps out, using the other wrecks for cover on approach to the radiation cage, pausing to eavesdrop on a murmured conversation before laying down to get a clear visual on the people welding a destroyer's hull into place.

He guesses that they nearly have it ready to fly, and wonders briefly if it might be plausible to get a couple of thousand out.

One voice amongst the many stands out, at first because of its authoritative tone, but then familiarity hits him.

Ennis breaks cover with his weapon raised, finding Keo in its scope before advancing, keeping the crosshairs trained on his heart.

"Keo!" snaps Ennis, startling the workers, welding and support all halting at once.

He enters the cage in a firing stance, planting his feet when he's close enough that he couldn't miss Keo's center mass.

"You people all need to leave now," says Ennis, "there's a helo at the northeast clearing waiting to take you off world."

People begin putting down their equipment and hurrying out, forming small groups to make the run to safety.

"We have more where they came from," Keo tells Ennis.

Apple scans the faces when they run past, grabbing one guy by the arm, asking, "Do you know Deborah Zenith? Deb Barclay?"

"Deb? She escaped already," he tells her.

"Which way?"

He watches the others fleeing, getting antsy. "Her and Eli went to find the northern tunnel while we distracted the soldiers. They were supposed to come back when they found a way out." He shrugs. "That was a while ago." The guy breaks her grip, running after the others.

Nodding at the patchworked destroyer, Ennis asks Keo, "Can it fly?"

Keo looks back at the hulking ship, pointing up, at half click-wide barn doors on the cavernous natural ceiling.

"Sure. But you'd still have to open those. We were about to begin sealing off the hull. Another half hour and this beast is spaceworthy."

"I guess only you know the hashcode for the exit."

"No, me and my guys . . . they're all quite loyal though."

"If you open it, we'll take you and them with us."

"They'll die here too. We all made a pact."

"Hah. Let's see if that whole deal comes to fruition. What do you want here, man? Why are you always so fucking obtuse?"

"I want you to lower your weapon, back out of here, and take your friends off my goddamn property sir."

"Keo, in a few minutes I'm going to let another few thousand people in here and jam them all aboard that ship."

Keo laughs. "Okay."

"Dude you can either come with us, or stay here on Earth and die. But we're not leaving without your workforce, as many healthy people as we can carry."

Still smiling, Keo asks, "How are you going to decide who stays, and who goes? Those people outside will crush each other to get in here."

Ennis doesn't have an answer.

"Why are you even here, Ennis? You'll still be a few hundred thousand rescues short."

"A few thousand closer."

Keo whips out the pistol with an imperceptible wave of the arm that Ennis doesn't catch until it's already popping at him. Even when he knows what it is, for just an instant, Ennis almost feels like he's waiting to see what will happen.

Two small rounds ping him in the chestplate while a third misses, Keo faltering when he notices that the shots haven't even broken Ennis's gaze.

While Keo steadies for the headshot, Ennis puts a .457 NATO round in each shoulder, a soft double tap that sends him reeling back one way then another--arms pellet drumming, spinning face down onto the floor, able now to operate neither arm nor weapon.

Ennis slides a foot closer, nudging Keo's leg.

"I'm awake you dickhead." Keo groans, rolling over. "Just fuck off, Ennis. You ruined it for everyone you shit. They wouldn't have gotten away with half of what they did, if not for Amsterdam. In the end they only locked us all down even tighter, because they could really wave this one in people's faces." Blood runs down Keo's neck, pooling in his dark hair and around the shredded shoulders. "Just do it then. Fucking do it you cunt."

Ennis now lowers the rifle, sitting down there on the dirty floor so that he can lean over Keo, so they can see eye to eye.

He tells Keo, "Without death you're not alive."

And Keo laughs at that, so hard that Apple and Weaver hear it as a scream, emerging from cover briskly but with caution, awed at first by the size of their surroundings, then aware of one figure crouching over another in the distance, below the centerpiece Ombudsman destroyer.

Coming within earshot, watching Keo and Ennis in their almost embrace, Weaver is feeling an intense sense of déjà vu while Apple is sure that she's about to see a murder.

"It all ended with Imogen anyway." It's the first time that he is speaking her name aloud since the day he died.

"Otis's Imogen?"

Standing over the both of them, Apple asks, "What about her?"

Keo looks straight up, at rock, moss and metal. The last three people he'll ever see, and he nothings them. That's the really funny part.

Then he tries to think of three people he actually would want to see here, three appropriately siginifcant personalities to appear at his side when facing the end of times as well as self. He can only really picture Otis, with the proviso that it's complicated.

"I put a fucking gun to her head . . . and I thought, I can't kill this poor lady. But I knew that if I didn't, I'd be in exactly her position inside twenty four hours. I saw myself kneeling there like her, begging someone else to stop. I saw my own tears. I'd cry like a little bitch when my moment came. When this moment . . . I would fucking beg. I'd . . . well. That's a bit ironic I guess. Now that we're here, we all know it's going to happen but it doesn't really seem to matter so much. You'll shoot me in the heart, the head, whatever. And it will all just stop."

Keo laughs again, quieter with shock wearing off, pain restricting even his facial movement.'

Ennis glances up at Apple, who prods, deadpan, "So you killed Imogen?"

Keo grins involuntarly, neck veins twitching, eyes glazed over with tears and a thousand yard stare. Ennis sinks into his own arms, truly beaten by this of all news. Only Apple and Weaver are distant, both having experienced enough grief that their ears always ring with it.

Keo watches the rifle hang from its strap while Ennis sits quietly, knowing he will never lift it. He can feel their contempt. And he sees that it's soon. He keeps telling himself not to panic. He is free. The instant it happens he will see and feel a spiral of light followed by pulses of the Godhead entering through his crown circle to bring him aboard the Impossible Wheel.

If his understanding of the Lunar texts is correct, the next three minutes will be eternity to his dying mind.

Weaver places a hand on Ennis, helps him stand, then ushers he and Apple back out through the cage door, closing it behind them.

Keo hears them all go, sobbing louder when they're through the vent seals, his pained cries and thrusts echoing up the tunnels even while they surface.

outlook

Otis can see the majority of Keo's mercenaries concentrated in encampments around the base of the ninety floor control tower, going in and out of the building while patrols form up and roll out, others returning in convoys of six vehicles who by the look of them have spent twelve hours hunting and cutting down refugees bold enough to scale the border fences.

Running at a crouch, Otis darts behind the last supply truck and hops its rear bumper, falling almost short due to his weaker leg, scrambling onto the roof to lay flat while it passes through assembled groups of mercs who chat together and eat standing up.

He ascends rapidly in one of the three lifts, passing seventy. Expecting unknown levels of resistance, he has a pistol loaded with safety off and the rifle still on his back but ready to roll. When the doors open and he peers out, scanning the upper viewing deck for suitable cover, Otis discovers that he's been overthinking it. A huddle has formed at the northeast set of windows, Otis joining all of Keo's men there to watch over their shoulders as a fresh tidal wave approaches the dockyards and refugees, still seventy clicks out, cartoonishly large.

Unnoticed by the mercs, Otis looks around for the panel that controls gates and accessways. Finding the terminal, he leans forward over the workstation to watch the main three barrier gates roll open while the crowd below surges, roaring their approval.

Finally the soldiers in the room catch sight of the crowd filling the laneways and access roads below, screaming down the ramps that lead to rumored potential of escape. The soldiers all watch Otis back out of the room and re-enter the lift, no one really feeling the impulse to try stopping him, given the gravity of the moment.

At the foot of the tower Otis joins the surging crowd, running with their flow toward the tunnels, his attention drawn to the conspicuous mass of three people pushing against the tide.

The reunion is short lived, Apple explaining that the destroyer should be capble of flight by the time they get back down there. Otis tells them about the wave, everyone painfully aware that there's not enough time to make it to their planned rendezvous with Umberto.

Their best bet, says Otis, is to get to higher ground for the initial tidal surge, then make their way across water to the extraction.

command centre

While the others guide people down to the radiation cage Weaver stands on the destroyer's bridge with three refugees who claim to have had experience aboard naval vessels.

Weaver has never been a part of the military, but spent a year and a half flying simulators as a hobby. Given the lack of confidence he finds in their eyes, he of course volunteers to pilot the ship out of its cage.

Before he falls back to the control tower, Ennis runs another sweep of the area for Keo, finding only an ominous bloody trail into the south tunnel, making a mental note to be wary of the man appearing out of nowhere at a crucial moment.

overwatch

The wave hits with a shuddering crash, becoming a roar as it consumes everything standing on the dockyards below, sweeping away the flimsier hangar and admin structures.

It floods the sections of tunnel that were unable to be sealed off, washing away the few million refugees remaining outside within the space of a breath.

Up in the tower, the foundation's shaking can be felt reverberating through the floor and furnishings, threatening to go at any moment but never truly dislodging from its deeper moorings. Below its sixtieth floor though, the powerful wave has taken all but the frame.

powered flight

In the radiation cage, two and a half thousand people have managed to cram into the destroyer, Weaver giving the order to seal the outer hatch and prepare for launch while those who could not fit aboard seek out tunnels that will seal.

The engines don't light right away, sending a wave of panic across the makeshift bridge, until the reassuring surge of power that staggers them in unison as the big ship finally jumps, then steadies. A meter off the deck, Weaver brings it around within the big space to give his side gunners a better look at the ceiling, slowly raising the destroyer's mass to lift its belly over the top of the cage, into clearer space.

They open up on the blast doors with heavy artillery, collapsing them almost instantly thanks to the water's weight, Weaver waiting several minutes for the debris to drop before even attempting to maneuver beneath the hole. A three sixty degree waterfall emerges from the dust and steam, beginning to flood the entire hangar below.

When he has enough confidence that the larger pieces have fallen, Weaver yaws into the stream and brings it up through the ceiling, then into open sky.

sendoff

Ennis, Apple and Otis stand on the tower's roof, watching the destroyer emerge from below the waterline, drifting upward for a ways before gaining momentum, taking no chances, with a fast break for orbit.

Otis is first to hear the missile batteries atop the tower's defense rig swing around, pointing them out to the others. Flagging the weapon terminals he saw three floors below, Otis is already taking stairs six, seven at once. Ennis is hot on his heels fumbling with a rifle, seeing that Otis has neglected to even arm himself, becoming decoy guy in any potential firefight by default.

A floor to go, and they hear the missiles launch. Still a chance they can send an early detonate. Assuming the firmware is up to date, they have twenty seconds tops.

On the eighty-seventh floor Ennis's current worst fear is realized when he comes upon Keo shooting at Otis, struggling to aim even a compact shotgun with his injuries.

"Keo you fucker, there are two thousand people aboard!"

Ennis shoots at Keo, but for a chubby crippled guy he's swift, darting behind the terminal's alloy projection drivers, aware that they wouldn't dare shoot out lenses needed to interface with its weapons.

On the other side at its controls, Otis is flicking up menus, familiarizing himself with about eight years' worth of weapon guidance advances in one stilted, self taught crash course.

"I'm locked out," he tells Ennis with almost hysterical terror in his throat.

Ennis fires a set of three at the wall behind Keo's cover, screaming. "Weaver's aboard, Keo, send the abort!"

"Okay I'll do it," says Keo, coming out, arms limp at his sides, the weapon tossed. Ennis traces his walk to Otis's side with the iron sight, both anxiously watching Keo lean over it.

"Reboot command client authorize K-V-Nine-Seven-Six," says Keo, too fast for them to interrupt, spitting in Otis's face before collapsing against the projectors. The whole set of displays disappear, flickering into a ninety second config sequence.

Otis kicks him square in the nuts, laying into his chest and face when he curls up, Ennis running over to feel along the terminal's smooth projection enclosure in the hope of finding a physical keyboard or override.

Coming up with nothing, shoulders sagging, the helplessness of their situation is crashing down on Ennis in waves, a complete realization of what they will see happen now. It dawns on them both like a light fading up.

As with all who have seen combat and survived, they are cut off from the impact now, a safety measure that will allow them to complete the mission before considering its gravity.

Otis then Ennis walk closer to the window, watching four lights streak almost past the destroyer's fleeting pilot lights, exploding in quick succession, a series of flashes then thumping, and from the resulting smoke a spiraling cloud of orange and blue light, the already delicate core of the falling ship going critical while it drops the last two hundred meters to churning surf below.

Its atomic stack continues to glow while it cuts through the surface with a significant splash and explosive steam cloud, the whole wreck disappearing through the foam with a final, brief flash.

Ennis leaves Otis at the window, returning to the half conscious Keo. Whispering, Ennis says, "If you thought living with Imogen was hard, this one is really going to fuck with your head. I know you only want me to kill you, because you're too scared to do it yourself. You no doubt think that what comes next will be bliss, or devoid of any conscience."

"I'll tell him," says Keo. "You have to do it or he'll know what I did and it'll just screw him up even more."

"In the afterlife," hisses Ennis, "we judge ourselves. Only you."

Keo has slowly started looking Ennis right in the eye, an intent scowl beginning to cross his face.

"You have no concept of what I know," says Keo, "and what you think you know has warped your understanding."

"No. I've read the Lunar Gnoses. I know you think a combat death will push you into the next circle of whatever shit you want to call it. But I've gone beyond knowledge into wisdom. I lived the crossing, went there and came back. More than once. I know because we've seen it."

Keo now falters, unsure. "And?"

"The texts got it ass-backwards." Ennis stands.

"Then what should I do?"

Ennis feels rage welling up, screaming at Keo, "You just killed the one guy who could have told you, dickhead!"

He slams the butt of his rifle down on Keo's head, hard enough that it doesn't feel harmless as intended.

He returns to Otis at the window to give him the tap, nodding toward the stairs.

They leave Keo there once again helpless but alive on the floor, meeting Apple at the stairwell door, all three silently descending the tower stairs.

worst day on Earth

When it becomes apparent that there's a twenty floor gap between them and the receeding waterline, still with no feed or comms coverage, the three return to the rooftop launchpad, Ennis and Apple sitting to catch their breath while Otis physically patches into the empty launcher's RF dish to broadcast an SOS template. Otis scans the horizon, trying to spot the Jughead. "I hope he saw the wave coming. If he even stepped outside to take a leak, and couldn't get airborne in time . . . "

The tower shudders again, a fresh round of quakes collapsing the skeletal tops of buildings in the distance. From one of the crumbling towers, Ennis spots something small lift off, squinting to make sure he's seeing it before opening his mouth. "Guys."

The Jughead approaches fast, flying to the edge of the tower roof, slowing to a gentle halt to hover at its edge. There is no sense of panic anymore. Everything that could go wrong, essentially has. Otis and Ennis help Apple up, before climbing aboard too.

Gaining alititude, they can see a triangle of approaching tsunamis, with no more visible signs of land beyond the pinprick tip of their tower, also about to be swallowed whole.

The world has suffered complete oceanic overturn, Umberto informs them. News feeds are reporting pockets of survivors here and there, nearby rescue vessels having ignored Ombudsman orders in droves after the devastation of First Pass.

But there are no more people alive on Earth, except perhaps those who made it underground. No one would expect them to survive long anyway. Officially, a billion and change are dead. The true toll will never be known.

Breaking orbit there is only cloud and blue. White curves sweeping through it all.

Aboard the Jughead there's a shared sense of awe, and dread. Apple at last looks away from the window, moving up front to join Umberto. She freezes at the cockpit's entrance, suddenly aware that Jess is nowhere aboard.

"Oh, Umberto! Did she get away somehow?"

He shakes his head.

She can say nothing else. She returns to the rear to sit between Ennis and Otis, laying her head on Ennis's shoulder.

With enough restraint that neither man can tell, she is balling her fucking eyes out.

crawl home

.Earth-Mars Corridor—Transit Space.2173.08.08.20:53

The hyperspace gates all thrown from alignment by Tyche's second pass, Umberto plots a three day sublight course that will take them to the far side of Mars.

With nothing left to do in the cockpit, he hauls himself through to the rear, stretching out his arms and legs while he floats between the others, coming in midway through a story about the island nation Amsterdam.

"Half an hour later they sealed the roads and bridges. They had gunships patrolling the moat so people couldn't swim across. All of the interior hospitals had collapsed with the blast, so people needed to be extracted and taken to Rotterdam or even Bruges, but Ombudsman couldn't put together an agreement with Healthcom or Worldsnet for a safe contamination corridor. After a brief debate they just sealed it all off. There's still footage floating around somewhere, of people being shot as they tried to cross the Hartog bridge. If you're thinking, no, that can't be right, I saw the Hartog bridge destroyed, that's because they pulled it after the fact, to take it off the table as an option for the few thousand survivors who made it out through a nuclear wasteland only to be turned away at the border. There was no plan, and orders were left purposefully vague. The peacekeepers were simply told, don't let people out."

Sipping unsweetened coffee from a silver bladder, it's finally dawning on Umberto that Ennis is Ennis Pryce.

"I guess it wouldn't have mattered. Even people outside the island were too badly burned to survive. They might have been treated for months watching their skin come off and hair fall out before dying in pain. What's the point? In the end Ombudsman probably did them all a favor, even the ones who spent hours without pain relief trapped inside the walls. Better that, than months of semilucidity."

Ennis trails off, becoming aware of Apple, Otis and Umberto. Then he says, "Weaver's with his girls."

end of the line

Umberto brings the Jarhead to a soft landing in wine-making country just outside Isidis's Granger Valley, a canyon system famous for its caves and vineyards. Taking as much as they can carry, Otis tosses in a magnesium shell to incinerate the Jarhead and remaining weapons, watching it slowly burn for hours in the distance before the fuel cells finally catch and explode, echoing through the valley.

blank stare

.Jarlsburg Ridge.OM/NP5002—Mars.2173.08.13.15:25

With Umberto there neither Otis or Ennis feel comfortable expressing a heartfelt goodbye to Apple, nor would they really know one were they to try. Instead they just silently watch the driver pack Apple and Umberto's bags onto the bus and then hustle back on board, obviously eager to make his run of stops before bedding down in Isidis.

She says nothing either.

Umberto nods then climbs the stairs, a brief smile from her with just the mouth and she's gone too.

The bus pulls away and descends into canyon roads, Ennis tracing its line through the desert with an occasional glimpse of rising dust while they climb foothills.

Otis checks the sky regularly for signs of aerial surveillance.

At the first small town or settlement, they will claim to have been robbed, and left in the desert. They will work for food and shelter, then after a few months pack up and start over somewhere else.

It's possible that Ombudsman has moved on, and is no longer interested in seeking their rendition. But to find out for certain will take time, and may cost them their lives regardless of Mailer's current disposition.

Their foreseeable future is not any kind of life, only an escape route. Potentially decades long and wide as space itself. They have both seen enough to know how this will go.

There is an unvocalised solidarity previously alien, a shared belief that death is preferable to living with the abandonment of integrity. For Otis, and Ennis, trust is no longer a liability.

About the Author

Rob Hackney is a Caucasian male of approximately average height and breadth. Since 2002 he has been writting for film, television, and comic books, as well as contributing concept art and storyboards to Australian film productions.

Made in the USA
Monee, IL
04 November 2026

67375793R00174